# HIMALAYA

## *Emergence and Evolution*

K S Valdiya

Jawaharlal Nehru Centre for
Advanced Scientific Research

Universities Press

**Universities Press (India) Limited**

*Registered Office*
3-5-819 Hyderguda, Hyderabad 500 029 (A.P.), India

*Distributed by*
**Orient Longman Limited**

*Registered Office*
3-6-272 Himayatnagar, Hyderabad 500 029 (A.P.), India

*Other Offices*
Bangalore / Bhopal / Bhubaneshwar / Chandigarh / Chennai
Ernakulam / Guwahati / Hyderabad / Jaipur / Kolkata
Lucknow / Mumbai / New Delhi / Patna



First Published 2001
ISBN 81 7371 397 9

*Typeset by*
OSDATA, Hyderabad 500 029

*Printed in India at*
Orion Printers, Hyderabad 500 004

*Published by*
Universities Press (India) Limited
3-5-819 Hyderguda, Hyderabad 500 029

*to*

Professor **C.N.R. Rao** FRS

who instilled in me
a new sense of
academic adventure

# Contents


*Foreword* vii

*Preface* ix

1. Himālaya—the Majestic and the Bountiful 1
2. India Collides with Asia 21
3. Water-Divide and Establishment of the Himalayan Drainage 27
4. Emergence of the Himādri as a Mountain 40
5. A Flashback into the Past 46
6. Birth and Development of the Śiwālik 58
7. Evolution of the Mountain Barrier and Onset of the Monsoon 68
8. Collapse of the Mountain Front and Formation of the Indo-Gangetic Basin 79
9. Evolution of the Indo-Gangetic Plains 91
10. Tectonic Tumult Goes on 108

*Glossary* 117

*Appendix - 1* 133

*Appendix - 2* 134

*Selected Books on Geology and Geography of Himālaya* 135

*Index* 136

# *Foreword*

The Jawaharlal Nehru Centre for Advanced Scientific Research established by the Government of India in 1989 as part of the centenary celebration of Pandit Jawaharlal Nehru, has completed the first decade of its existence. Located in Bangalore, it functions in close academic collaboration with the Indian Institute of Science.

The Centre is an autonomous institution devoted to advanced scientific research. It promotes programmes in chosen frontier areas of science and engineering and supports workshops and symposia in these areas. It also has programmes to encourage young talent.

In addition to the above activities, the Centre has a programme of publishing high quality Educational Monographs written by leading scientists and engineers in the country addressed to students at the graduate and postgraduate levels, and the general research community. These are short accounts introducing the reader to interesting areas in science and engineering in an easy manner so that later study in greater depth and detail is facilitated.

This monograph is one of the series being brought out as part of the publication activities of the Centre. The Centre pays due attention to the choice of authors and subjects and style of presentation, to make these monographs attractive, interesting and useful to students as well as teachers. It is our hope that these publications will be received well both within and outside India.

V. KRISHNAN

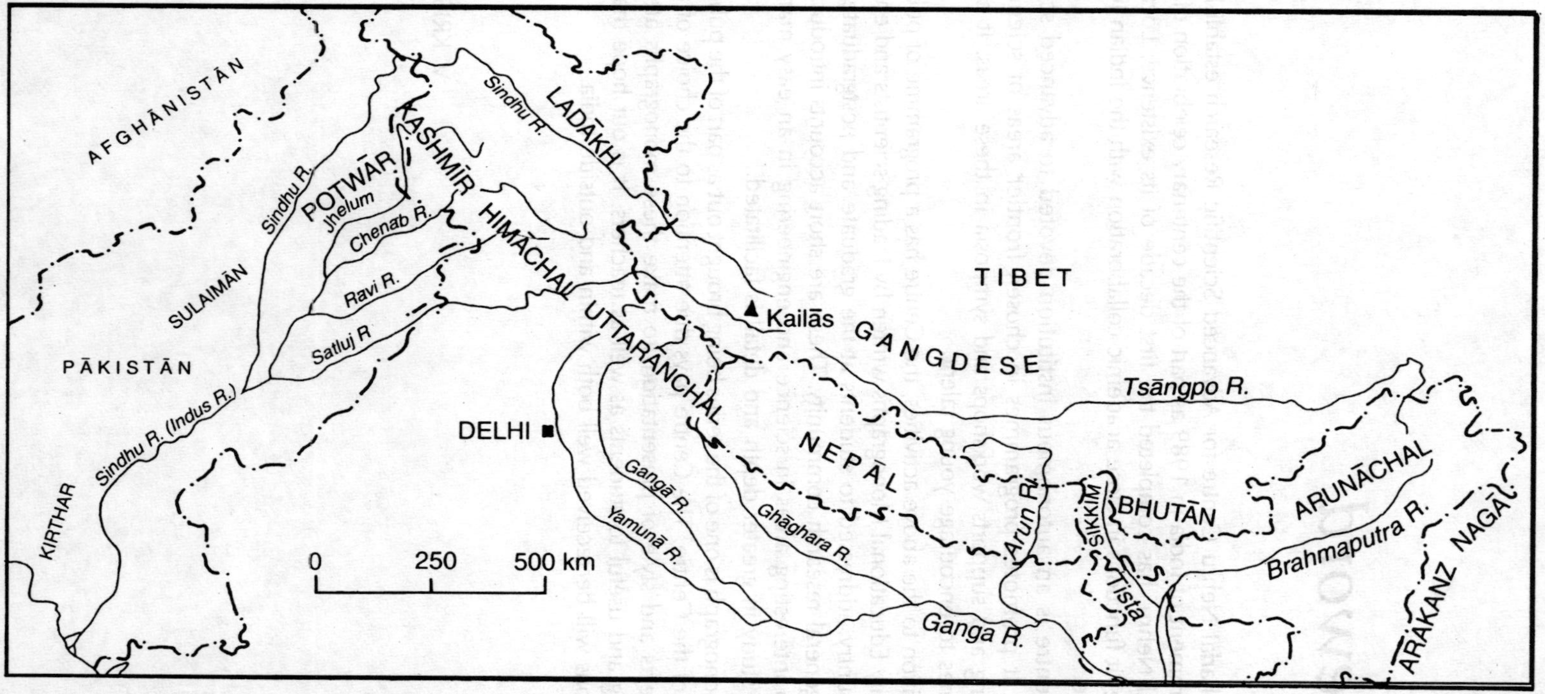


**Figure P1** Various geographical–geological sectors of the Himalaya.

# *Preface*

When I wrote *Dynamic Himālaya,* I had in mind readers who understood the basics of geology. I now realize that the fascination for the Himālaya runs far wider than I had imagined. This little book is written for those who are not familiar with geological jargon. Dispensing with the technical terms and the names that geologists use to describe rock-assemblages, this book provides a broad though brief and updated coverage of the history of the birth and development of the Himālaya. It is a simplified synthesis of geomorphological, geological and geophysical data, related to the pattern and sequence of events leading to the emergence of the world's highest but youngest mountain. Presented in the context of the wider panorama of the evolution of the Indian subcontinent, the book highlights the crucial developments that overtook the northward-moving landmass of India.

Although I have unabashedly projected my own perceptions based on the work I did in the central sector of the mountain arc, I have also drawn freely upon the works of a very large number of researchers.

I am grateful to my friends who have helped me bring out this book for the general reader.

K.S. VALDIYA

1

# *Himālaya—the Majestic and the Bountiful*

Guarding the northern frontier like a mighty sentinel, the Himālaya isolates the Indian subcontinent securely from the rest of Eurasia (Figure 1.1A). It is the provider of life-support systems for six countries. Not only does the Himālaya control the climate of Asia, it has also moulded the lifestyles of the peoples who inhabit the lands in and around the mountain domain. Majestic and many-splendoured, the Himālaya is a colossus among the great mountains of the world. The great poet Kalidas aptly described it as the king of mountains:

अस्त्युत्तरस्यां दिशि देवात्मा हिमालयो नाम नगाधिराजः

"In the north lies the divine-spirited king of mountains, named Himālaya."

The "king of mountains" is a part of the great chain of mountains that stretches from Spain in the west to Indonesia in the southeast, embracing the Baltics, the Carpathians, the Zagros, the Hindukush, the Himālaya and the Patkai-Arakan Yoma. The geologists include the Kirthar-Sulaiman ranges of Pakistan in the 5 200 km-long mountain province of the Himālaya.

The Himālaya is among the youngest mountains of the world, and it is still undergoing structural changes and growing (Figure 1.1B). This is evident from the increasing heights of its peaks. The tectonic turmoil of growing up is manifest in the upheavals occurring in the giant frame and in the spells of twitching and quivering into which the Himālaya is frequently thrown.

The world's highest and youngest mountain massif was formed as a result of the coming together, and eventual collision, of India and mainland Asia. The stupendous pile of sedimentary rocks lying in the frontal part of the Indian subcontinent was compressed, squeezed and wrenched into a giant edifice called the Himālaya. As the northward-moving India pressed against Asia, mountain ranges were moulded around the projecting promontories of the Indian landmass. A very remarkable feature is the festoons of curved mountain ranges that make spectacular knee-bends at the northwestern and northeastern corners (Figure 1.1A). In Kashmir the whole of the mountain system abruptly turns southward, making an acute angle near the pivotal point. In the northeast, the mountain ranges bend around the Namcha Barwa knot.

(A)

(B)

**Figure 1.1** (A) The high mountain rampart guards the Indian subcontinent and isolates it securely from the rest of Eurasia. (Based on a 1969 portrayal by Bruce Heezen and Marie Tharpe). (B) Even as the giant frame of the Himalaya is undergoing structural changes, its peaks are growing in height. The Shivaling peak in the Gangotri area from where the Ganga emanates.

## Physiographic Domains

The 2 400 km-long and 300 to 400 km-wide expanse of the Himalayan province is divisible into four physiographic domains. Each of these domains has its distinctive geological identity, its peculiar structural architecture, its own distinctive assemblage of rocks, and contrasted physiographic layout. Recognized as terranes, the Himalayan domains are known as the *Śiwālik* in the south, the *Lesser Himālaya* in the middle, the *Himādri* or *Great Himālaya* further north, and the *Tethys Himālaya* in the far north (Figures 1.2 and 1.3).

The *Śiwālik terrane* in the south is separated from the Indo-Gangetic Plains by a series of reverse faults. These faults are concealed under a discontinuous apron of gravelly debris that slipped down the hillsides, and of detritus deposited by mountain rivers at the points of their emerging into the plains. This apron of debris is described as *Bhabhar* in the central sector of the mountain arc. The 250 m- to 800 m-high Śiwālik ranges are generally very rugged (Figures 1.2 and 1.4A), but west of the Gaula River and east of the Sharada River there are flat stretches called *duns* within the hills. The duns are intermontane synclinal valleys filled with gravelly deposits dumped by rivers in their wide channels. The Śiwālik terrane is composed of sedimentary rocks made up of material deposited by rivers in their channels and floodplains, about 18 to 1 million years ago. The Śiwālik Hills are covered by dense tropical–subtropical forests east of the Yamuna River. The terrane is, on the whole, sparsely populated east of the Ganga Valley, but harbours a rich variety of wildlife.

The outer ranges of the *Lesser Himālaya* abruptly rise to elevations of 2000 to 2500 m against the Śiwālik hills. In the northwest the PirPanjal-Dhauladhar Ranges are more than 3 500 m-high, and in southcentral Nepal the Mahabharat forms a 3 000 m and higher mountain rampart. In the central sector, stretching from Nepal through Uttaranchal to Himachal Pradesh, the middle belt of the Lesser Himālaya has comparatively mild relief, being characterized by a rather undulating landscape, rounded hilltops and gentler slopes (Figures 1.2 and 1.4B). Rivers and streams flow quietly without hurry, but rush through gorges where they cross recently faulted-up mountain blocks. The Lesser Himalayan terrane comprises a very thick succession of sedimentary and associated volcanic rocks, ranging in age from more than 1600 million years to about 540 million years, and overlain by sheets of metamorphic rocks of Precambrian antiquity and granites belonging to two periods—1 800 to 2 000 million years and 500 to 540 million years. Once thickly covered by forests but presently considerably destitute of the sylvan cover over the greater part, the Lesser Himālaya is the most populated terrane of the Himalayan province.

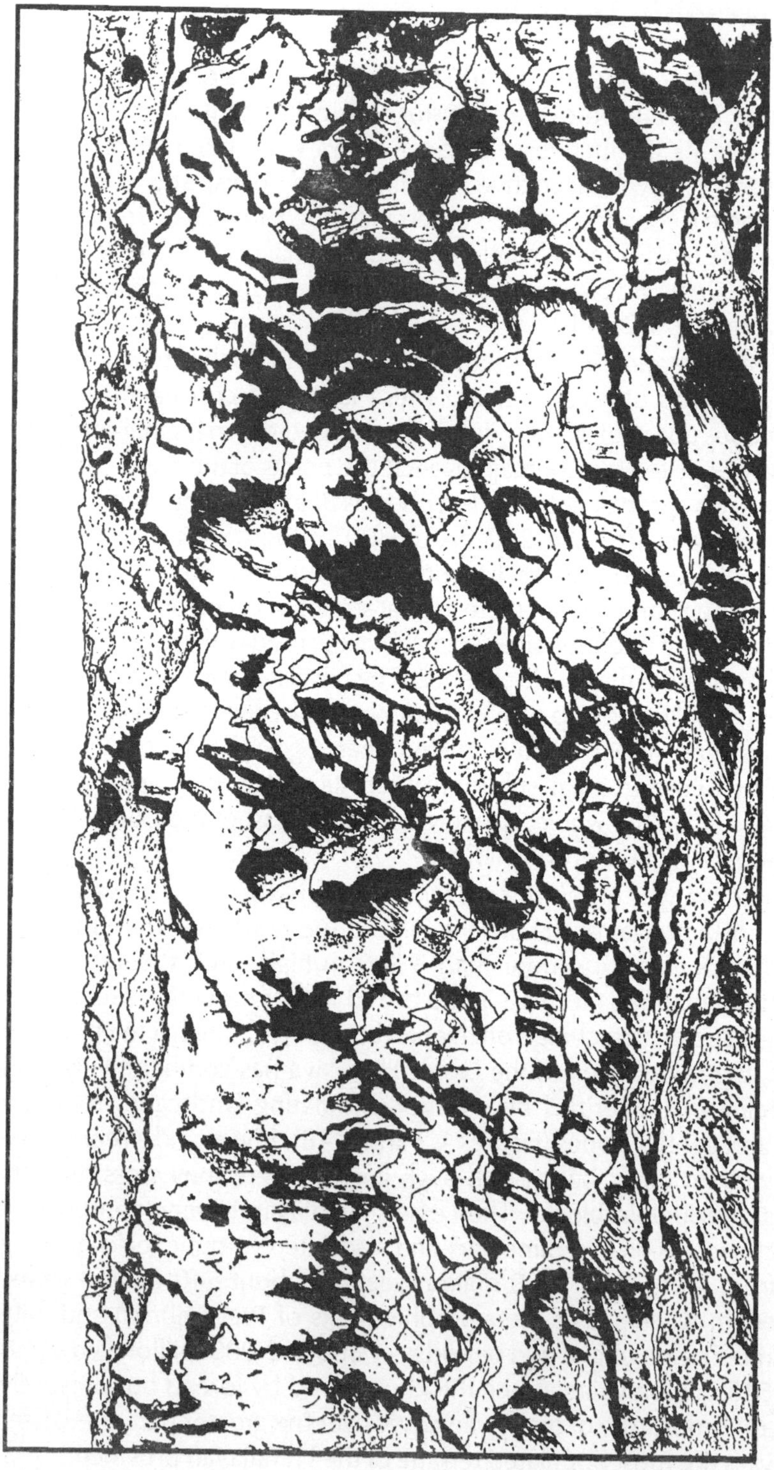

**Figure 1.2** A speculative portrayal by Sudip K Paul of the four physiographically and geologically distinctive domains or terranes of the Himalayan province. The outermost low-lying hills make up the *Śiwālik*. The *Lesser Himālaya* is represented by relatively gentler topography below the rugged high mountains of the *Himādri*. The high but gentler terrain in the north is the *Tethys* domain.

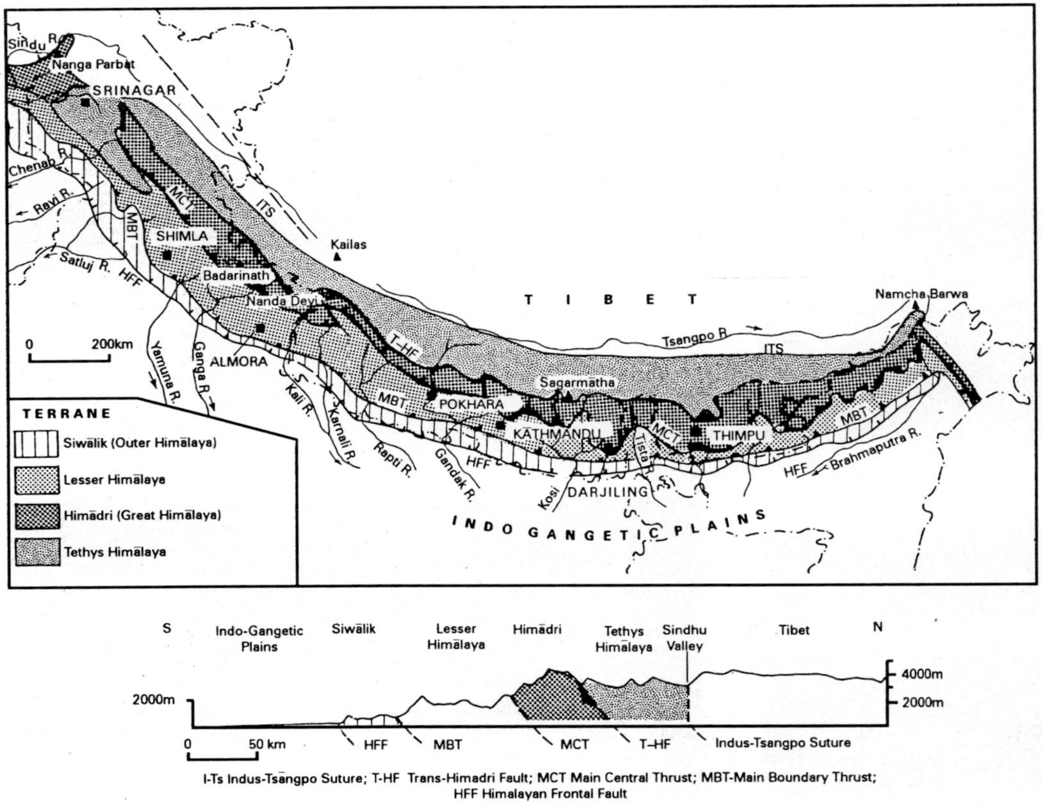


**Figure 1.3** Four physiographically distinctive and geologically contrasted domains or terranes of the Himālaya. Each of the terranes is delimited by deep, long faults. The Indus-Tsangpo Suture marks the junction of the Himālaya and mainland Asia.

**Figure 1.4** Glimpses of the topographic peculiarities of the four terranes:(A) Siwalik, (B) Lesser Himalaya, (C) Himadri, and (D) Tethys Himalaya

Overlooking the Lesser Himalayan terrane stands the *Himādri* or *Great Himālaya*, 3 000 to 8 000 m high. The southern face of the extremely rugged ranges of the Himādri is marked by high scarps that inspire both awe and reverence (Figures 1.2 and 1.4C). In the northwest is the 8 126 m-high Nanga Parbat; the central sector embraces Leopargial (6 791 m), Kedarnath (6 900 m), Badarinath (7 138 m), Nanda Devi (7 817 m), Dhaulagiri (8 172 m), Sagarmatha or Everest (8850 m) and Kanchanjangha (8 598 m); and in the extreme northeast, in solitary eminence, stands the 7 256 m-high Namcha Barwa. These ever-snowy spectacular ranges form what is known as the axis of the Himālaya — what the poet Kalidasa had described as "the measuring rod of the Earth" —

पूर्वापरौ तोयनिधिवगाह्य स्थितः पृथिव्या इव मानदण्डः

"Stretching from sea to sea, it (the Himālaya) lies like the measuring rod of the Earth".

Mountain torrents dash down the steep slopes and roaring rivers rush through deep gorges. Characterized by an utterly youthful and yet forbiddingly rugged topography, the Himādri is the abode of snow and ice. The terrane is made up of metamorphic rocks that were formed under very high temperatures and pressures, and of the granites that were emplaced 500 to 540 million years and 19 to 21 million years ago.

North of the Himādri lies the vast expanse of the *Tethys Himālaya*. A cold and desolate domain, the Tethys terrane is sparsely populated and the settlements are huddled around clusters of trees in valleys. The fantastically beautiful ice-sculptured ranges (Figure 1.4D) of the domain are made up of sedimentary rocks that range in age from more than 600 to about 65 million years.

The Himalayan province ends against the mainland Asia along the valleys of the Sindhu and Tsangpo rivers (Figure 1.5). These rivers occupy the 50 to 60 km- wide zone of collision of India with Asia. Lying 3 600 to 5 000 m above sea level, the collision zone comprises extremely compressed and sheared rocks cut by deep faults. The rocks of this zone once formed the floor of the ocean and occupied an oceanic trench that lay in front of the continent of Asia. Sixty to fifty million year-old volcanic rocks are inextricably mixed up with the deep-ocean sediments.

North of the Sindhu–Tsangpo valleys is the highlands of Tibet and Karakoram—an altogether different landmass belonging to mainland Asia. It is an undulating plateau, more than 5 000 m above sea level. The southern front of the Tibetan Plateau embodies 40 to 60 million year-old granites all along the border of Tibet, making up the Ladakh-Kailas-Gangdese Ranges.

(A)

(B)

**Figure 1.5** Rivers Sindhu (A) and Tsangpo (B) occupy the zone of collision of the Indian and Asian continents. *(Photos: M E Brookfield)*

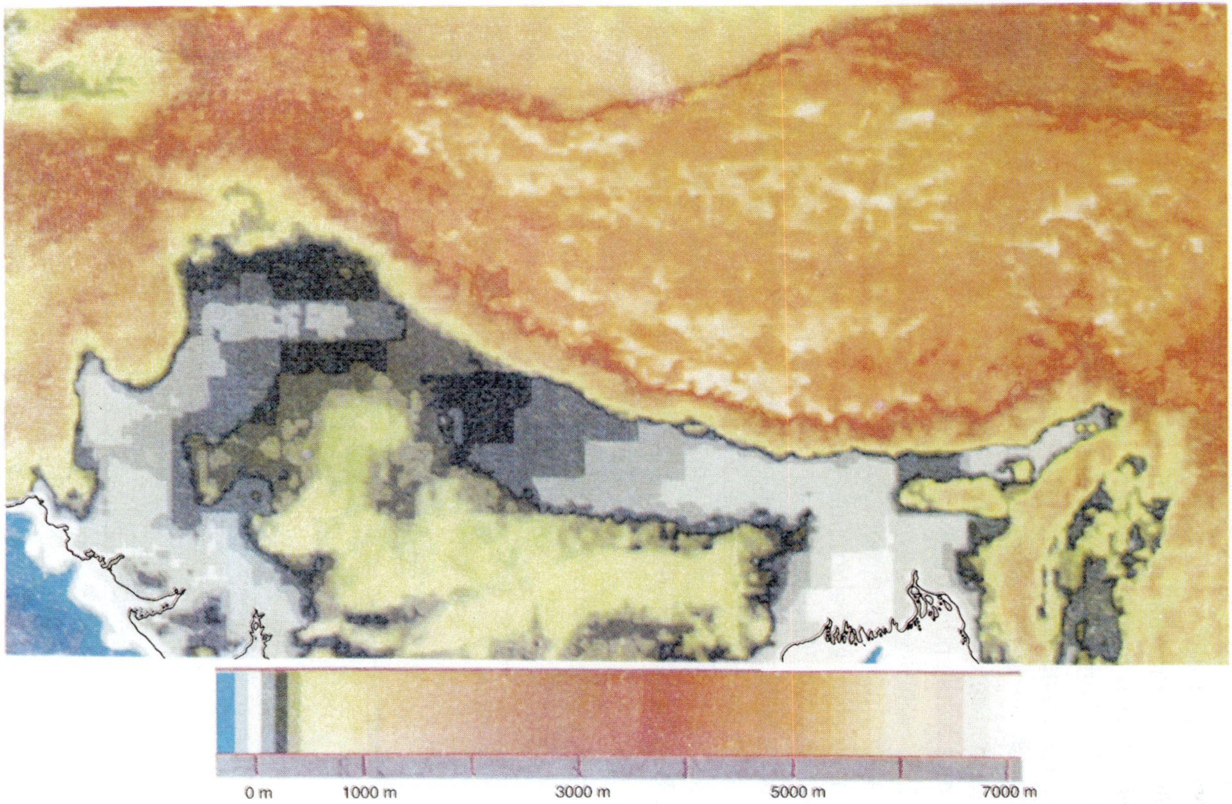


**Figure 1.6** The high Himalayan barrier causes precipitation of moisture of the north-flowing clouds in the form of rain and snow and stops the cold winter wind from Siberia entering into the Indian subcontinent.

## Controller of Climate

The 2 400 km-long mountain barrier sprawling 300 to 400 km wide across the northern edge of India and rising 500 to 800 m high above sea level, controls the atmospheric circulation over the continent of Asia. The lofty ranges cause precipitation of moisture from the clouds in the form of rain and snow. The monsoon clouds blowing in from the Indian Ocean are prevented from going northeast (Figure 1.6), culminating in the development of dry conditions and deserts in Ladakh and Tibet in the north, and in western Rajasthan in the south. The Siberian cold winter wind is likewise stopped from entering the Indian landmass, with the result that winters in northern India are less severe than they would have been if there had been no Himalayan barrier. The Himālaya thus exercises a moderating influence on the temperature and humidity over the Indian subcontinent.

It was the rise of the Himālaya above a critical elevation that brought about the advent and intensification of the monsoon climate. The unique cycle of six seasons — *vasant, grishma, varsha, shishir, sharad and hemant* — is the characteristic of the monsoon that sweeps the Indian subcontinent. The configuration and altitudinal peculiarities of the mountain ranges of the Himālaya are responsible for the variations of climate within the mountain province itself. Take the case of rainfall. If the high southern front of the mountains receives an annual rainfall of 240 cm/yr, the precipitation in the middle belt of the Lesser Himālaya is 150 cm/yr, while the south-facing slopes of the Himādri are washed by rainfall in excess of 240 to 350

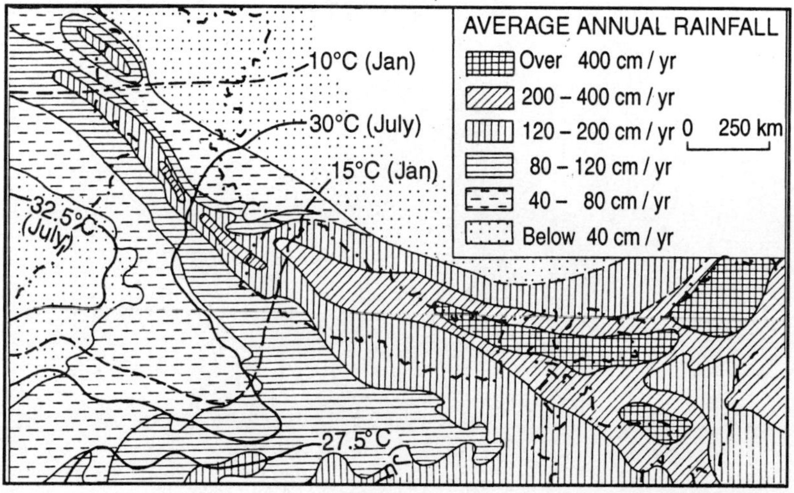


**Figure 1.7** Annual rainfall and January and July temperatures in the Himalayan province and adjoining Indo-Gangetic Plains.

cm/yr; and it is just 10–15 cm/yr across the Great Himālaya in the Tethyan belt (Figure 1.7). The eastern Himālaya experiences a higher rainfall than the northwestern Himālaya. The rainfall is 300 cm/yr at Darjiling, 50 cm/yr at Shimla in Himachal Pradesh, and a mere 10 cm/yr at Leh in Ladakh. The temperature similarly varies from belt to belt and sector to sector.

## Rich Biodiversity

The biodiversity of the Himalayan province is extraordinarily high. The eastern part of the Himālaya embraces the zones of the highest biological diversity, with more than 4 000 species of vascular plants per 10 000 km$^2$ area (Figure 1.8). This is exceptionally high even in the global context. In the larger part of the Himalayan province, the vascular plant species is above 3 000 per 10000 km$^2$ area.

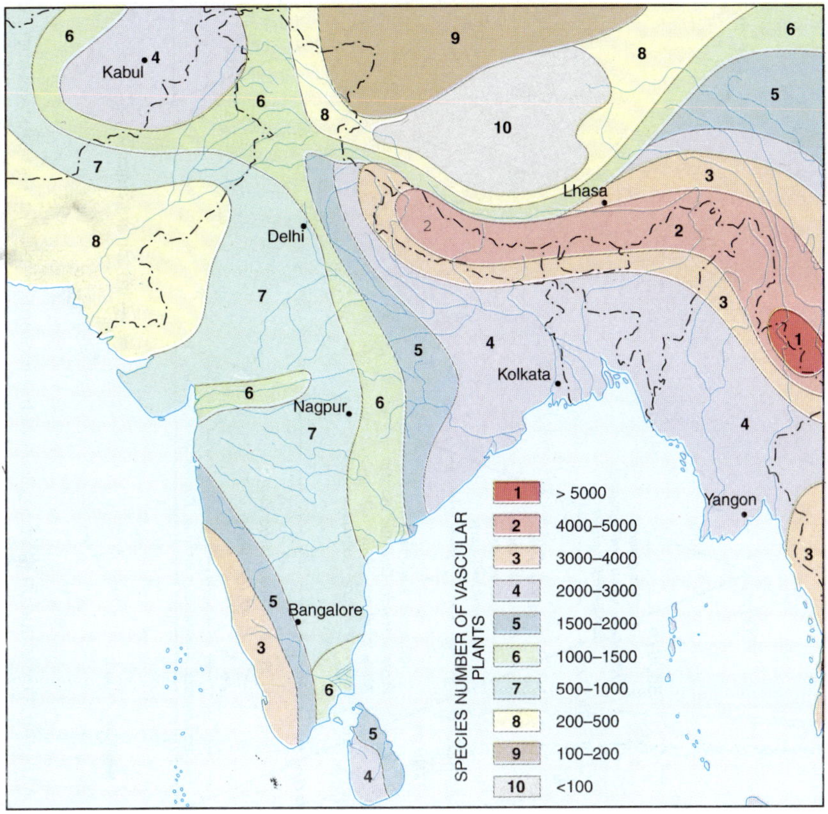


**Figure 1.8** Biodiversity zones of central Asia, embracing the Himalayan province and the subcontinent of India. (Based on a map in *Mountains of the World* edited by B Messerli and J D Ives, 1997).

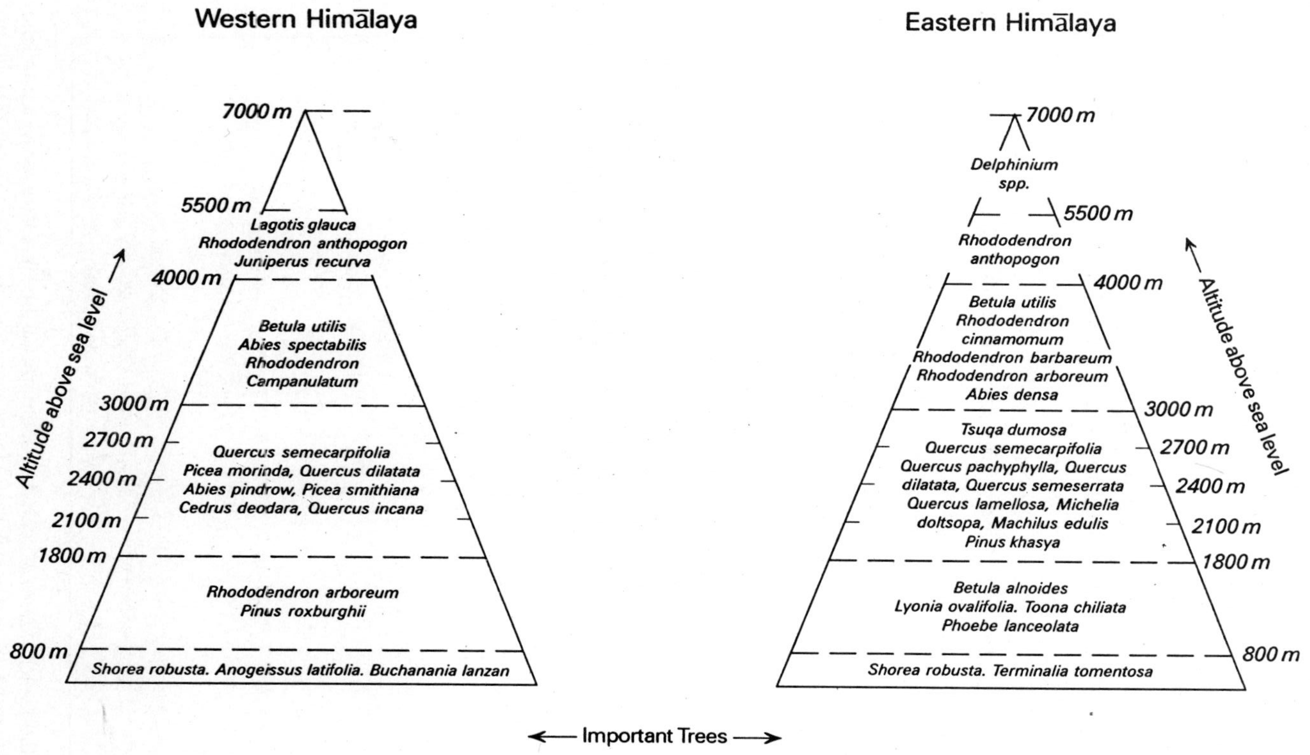


**Figure 1.9** Altitudinal distribution of vegetation of the Eastern Himālaya and the Western Himālaya highlights the compositional diversity along the mountain province.

Nearly 30% of the forest plants are uniquely Himalayan, and are not found anywhere else in the world. These include certain species of oak, rhododendron and pine. About 125 plant species are wild relatives of cultivated plants, including cereals, legumes and nuts. These constitute valuable gene pools which can be used in future for the improvement of crops. Of the 1500 species of angiosperms known in India, 20 to 49% occur in the Himalayan province. More than 30% of these angiosperm plants are entirely endemic and economically quite useful. Nearly 7020 species or 54% of the Indian fungi have been reported from this mountain domain. About 1159 out of the total 1948 Indian lichens occur in the Himalayan world. Likewise, 23% or 115 species of Indian bryophytes grow on the Himalayan slopes.

The composition of plant assemblages of the wetter eastern Himālaya is very different from that of the relatively drier western Himālaya (Figure 1.9). Even within the western sector, the forest displays a tremendous diversity of floral characters—from the dense evergreen tropical vegetation of the torrid Bhabhar and Śiwālik belts, through mixed deciduous trees with savannah grasslands in the salubrious middle mountain, to the sparse arctic-type plants in the cold high mountain belt. In the western Himalayan domain the following floral assemblage is seen. The forests of the Bhabhar-Śiwālik terrane are dominated by 'saal' (*Shorea robusta*), 'khair' (*Acacia catechu*), 'sheesham' (*Dalbergia sissoo*), 'haldu' (*Adina cordifolia*), and 'sain' (*Terminalia tomentosa*), along with infinite varieties of shrubs and grasses. In the lower altitudes of the Lesser Himālaya, the 'chirpine' (*Pinus roxburghii*) grows luxuriously on dry slopes with poor soil condition, while forests of oaks (*Quercus leucotrichophora, Quercus incana*), rhododendron (*Rhododendron arboreum*), and alder (*Alnus nepalensis*) cover the moist slopes with good soils. The productivity of this vegetation is quite high—the dry biomass being as much as 20 tonnes per hectare per year. The higher altitudes of the Lesser Himalayan mountains have forests characterized by 'kharsu' oak (*Quercus semecarpifolia*), 'tilonj' (*Quercus dilata*), blue pine (*Pinus wallichii*), acer, etc. The Great Himalayan domain is covered with forests dominated by silver fir (*Abies pindrow*), 'bhojpatra' or birch (*Betula utilis*), fir (*Abies spectabilis*), stunted rhododendron (*R. companulatum*), and junipers (*Juniperus squamata, J. indica*). Still higher, the forests give way to alpine meadows which have an infinite variety of flowering grasses.

In the eastern sector the assemblage of important plants is as follows. The low altitude (< 800 m), tropical deciduous forests dominated by *Shorea robusta* give way to subtropical forests with *Schima-Castanopsis* in the range of 800 to 1 800 m. Ascending the slopes from 1 800 m to 3 000 m the forests, the composition changes from Lauraceae, through the *Quercus lamellosa*, *Quercus pachyphylla* to *Tsuqa dumosa–Quercus semecarpifolia* assemblage. In the altitudinal zone, 3 000 to 4 000 m, the forests are dominated by the *Abies densa–Betula utilis* assemblage. *Rhododendron anthopogon* occurs at the 4 000 to 5 500 m altitudes. And, above 5 500 m, grow species of *Delphinium*.

The Himalayan forests abound in a wide variety of animals— elephants, tigers and panthers in the Bhabhar-Śiwālik domain in the south, including rhinos in the Nepal foothills and bears in the higher mountains. There are many animals which are uniquely Himalayan, such as 'kankar' (*Mutiacus muntjak vaginalis*), 'ghural' (*Nemorhaedus goral goral*) and 'bharal' (*Pseudois nayaur*) in the middle mountains; the snow leopard (*Panthera unica*), Himalayan black bear (*Selenarctos tibetanus*), musk deer or 'kastura' (*Moschus moschiferus*), and Himalayan tahr (*Hemitragus jemlahius*) in the higher realm. The bird life is fabulously rich—230 species have been spotted in the Kumaun Himālaya alone.

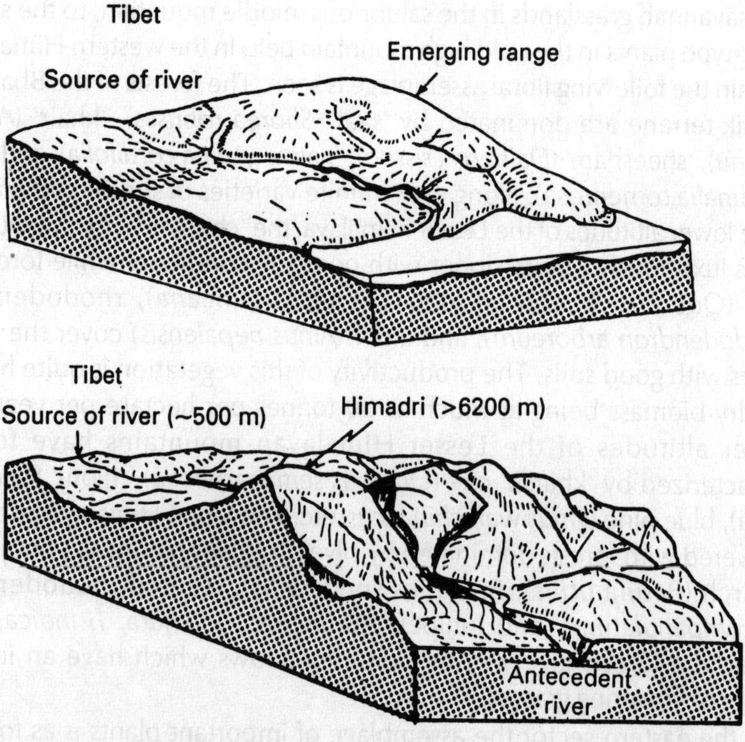


**Figure 1.10** Rivers of the Himalayan province. Significantly, the Sindhu, the Satluj, the Karnali, and the Tsangpo — among many others — arise north of the stupendous mountain barrier of the Himālaya in the region of the Mansarovar lake.

## Rivers and the Water Asset

The four great rivers of the Himalayan province spring from one region—the Kailas-Mansarovar lake tract in southwestern Tibet. The source area lies at an altitude of about 5 000 m above sea level, across the 7 000–8 000 m-high Himalayan barrier. The Sindhu flows northwestward, the Satluj goes southwest, the Karnali (Ghaghara in the plain) takes a southerly course, and the Tsangpo (Brahmaputra in Assam) flows east (Figure 1.10). These rivers flow through the channels they carved out at the outset—well before the Hima¯laya rose up as a stupendous barrier. Interestingly, the ***Matsya Purana*** describes the descent of the *Divyaganga* river near Bindusar lake nestling between the Kailas, Mainak and Hiranyashringa mountains, and taking three different paths as *Tripathaga* (going along three paths). The rivers are doubtless older than the mountain ranges they cross. They rush through deep gorges and chasms that have walls as high as 3 000 to 4 000 m, while the river bed is not higher than 900 m to 1 200 m above sea level. This wondrous spectacle of rivers rushing through deep gorges is best seen west of the Nanga Parbat and east of the Namcha Barwa.

Nearly 1 20 00 000 million cubic metres of water, on an average, annually flows down the rivers of the Himālaya. This water not only sustains and nurtures the teeming millions of the northern Indian subcontinent, but also has an enormous potential for irrigation and power

**Table 1.1** Annual average flow and potentials of the Himalayan Rivers.

| River | Flow (billion $m^3$/yr) | Irrigation Potential (billion $m^3$/yr) | Power Potential (megawatt at 60% load factor) |
|---|---|---|---|
| *Brahmaputra System* | 479.00 | 12.3 | 9988 |
| Brahmaputra | 455.00 | | |
| Manas | 47.10 | | |
| *Ganga System* | 459.84 | 185.0 | 11579 |
| Ganga | 23.90 | | |
| Yamuna | 12.30 | | |
| Gandak | 53.20 | | |
| Tista | 19.40 | | |
| *Sindhu System* | 207.80 | 49.3 | 6582 |
| Sindhu | 73.30 | | |
| Jhelum | 27.89 | | |
| Chenab | 29.00 | | |
| Ravi | 8.00 | | |
| Beas | 15.80 | | |
| Satluj | 16.66 | | |
| Total | 1147.00 | 246.6 | 28150 |

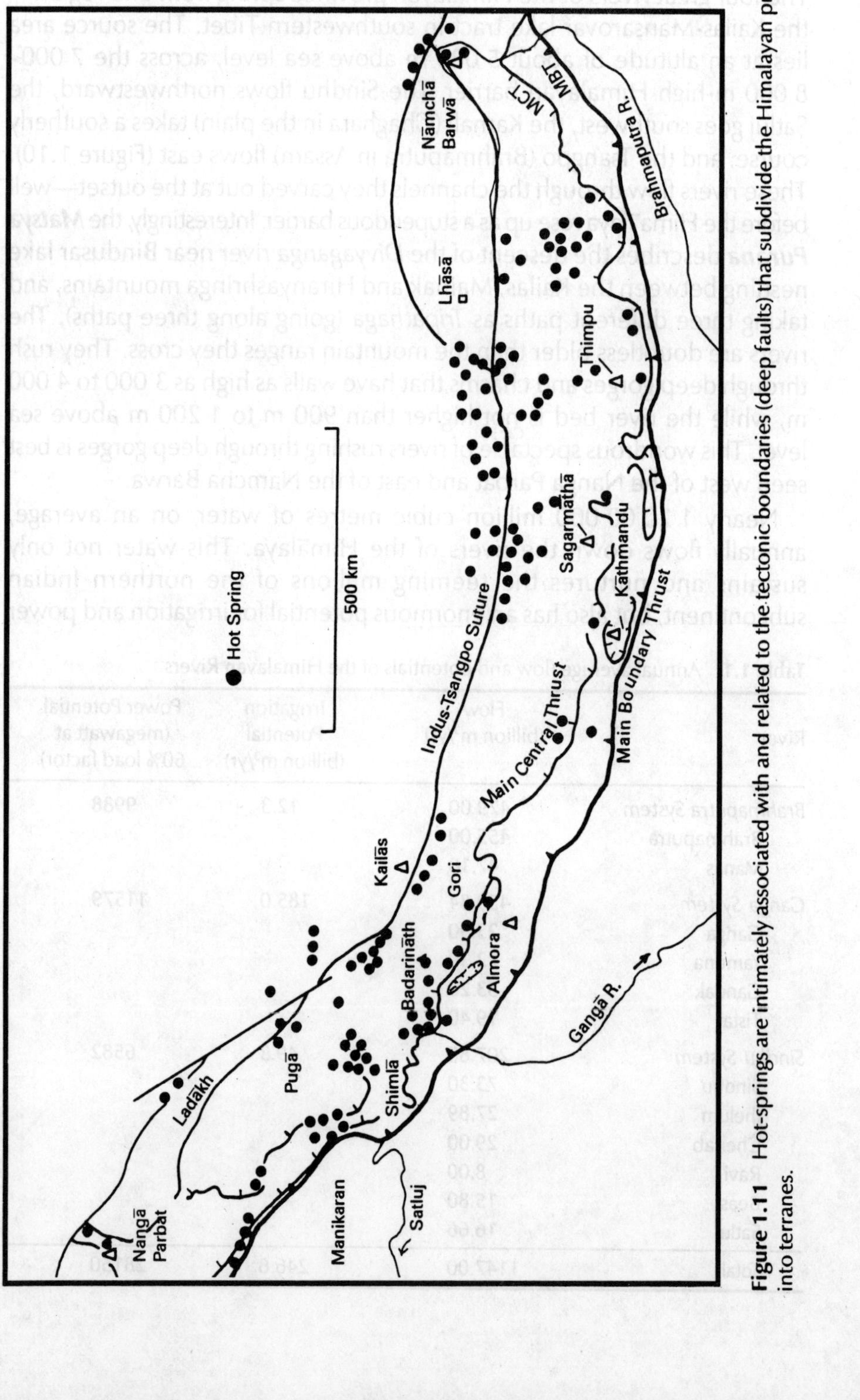


**Figure 1.11** Hot-springs are intimately associated with and related to the tectonic boundaries (deep faults) that subdivide the Himalayan province into terranes.

generation. One assessment shows that 2 46 000 million cubic metres of water can be utilized for irrigation and nearly 28 000 megawatts of electricity can be generated, if suitably tapped (Table 1.1). This is, no doubt, why the Ganga is described in the ***Puranas*** as *Jahnvi*—the life-giving river.

There are hot-springs throughout the Himālaya, occurring in the belts of tectonic boundaries that separate the four geomorphical–geological terranes. There is a particular concentration of thermal fields in the zone of the junction of the Himālaya and mainland Asia, along which flow the rivers Sindhu and Tsangpo. Yet another belt of high heatflow and large number of thermal springs, is related to the tectonic boundary at the base of the Himādri. There are 34 hot-springs in Ladakh, 34 in Himachal Pradesh, 37 in Uttaranchal, 7 in Sikkim and 11 in Arunachal Pradesh (Figure 1.11). These hot-springs have an energy-generating potential varying from 130 $mW/m^2$ to $468/mW/m^2$. Also, the hot water and steam can be utilized to heat buildings and glass-houses for growing vegetables and flowers, as is being successfully done in Ladakh.

## Mineral Wealth

Mineral deposits in the Himalayan rocks (Figure 1.12), particularly of the Lesser Himālaya and Śiwālik terranes, are considerable in magnitude and are either being mined profitably or have the potential of economically

**Table 1.2** Mineral wealth of the Himālaya within the Indian territory.

| Mineral | Proven Reserves in 1980 | |
|---|---|---|
| Limestone | 458 | m.t |
| Dolomite | 94 | m.t |
| Magnesite | 82.2 | m.t |
| Gypsum | 66.7 | m.t |
| Graphite | 26.7 | m.t |
| Lignite | 21.7 | m.t |
| Phosphorite | 18.1 | m.t |
| Bauxite | 13.6 | m.t |
| Coal | 11.6 | m.t |
| Rock Salt | 8.0 | m.t |
| Copper-Lead-Zinc | 2.2 | m.t |
| Steatite-Talc | 1.9 | m.t |
| Fluorite | 86 000 | t |
| Bentonite | 40 000 | t |
| Sulphur | 2 01 000 | t |
| Barytes | 13 200 | t |
| Antimony | 10 588 | t |
| Borax | 5 423 | t |
| Uranium minerals | Appreciable | |

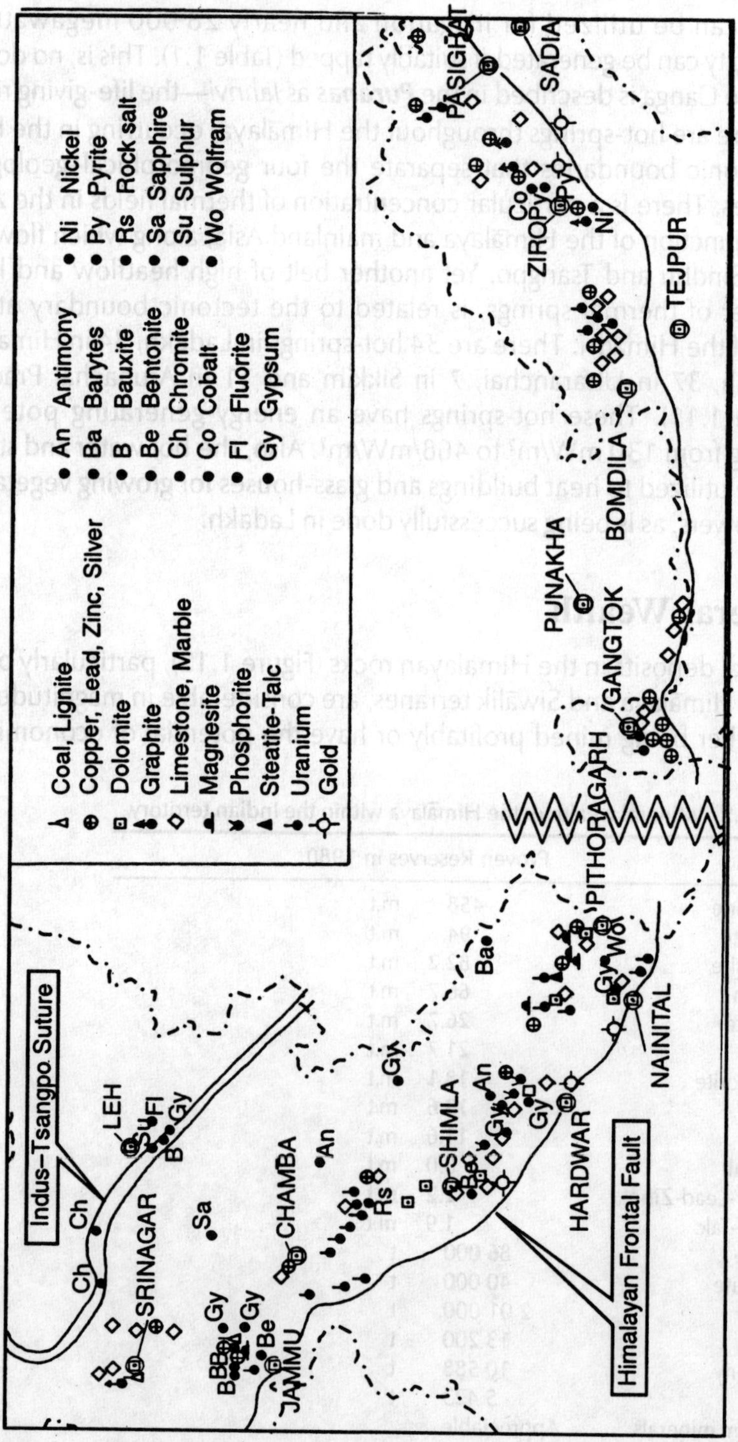


**Figure 1.12** Deposits of some important minerals in the Indian part of the Himālaya.

viable mining in the future (Table 1.2). Reserves of magnesite (carbonate of Mg), dolomite (double carbonate of Mg and Ca), cement-grade limestone (carbonate of Ca), gypsum (hydrated sulphate of Ca), roofing slate and paving-stone are very large and quite promising. Reserves of talc, phosphorite, lignite, rock salt, and base metals (primarily sulphides of Cu, Pb, Zn, etc.) are substantial. Very promising deposits of uranium-bearing minerals occur in many places in all the four terranes, but mainly in the Lesser Himālaya and the Śiwālik. These can be mined in times of emergency.

## Moulder of Lifestyles and Destiny

The Himālaya has witnessed the making of the history of the Indian civilization and the stirring events that shaped the destiny of the peoples of the subcontinent. It has moulded the lifestyles of the inhabitants of the lands in and around it. No wonder the *Nagādhirāj* evokes both awe and admiration. The snow-capped mountains of immense splendour and majesty, the roaring and foaming rivers, and the whispering forests and the smiling fields have attracted saints and pilgrims, poets and philosophers, adventurers and mountaineers, and warriors and fugitives ever since the "khas" people set foot there more than 2 500 yr B.P. Many of the later visitors also made the Himalayan domain their home. Today there are more than 58 million people living in the habitable places, mostly in the relatively mild and congenial belt of the Lesser Himālaya.

The people of the Himālaya belong to three principal racial stocks (Figure 1.13A–C). The northern territory of the Tethys Himālaya, stretching from Ladakh in Kashmir to northern Kameng in Arunachal Pradesh, is the home of the tribes of Tibetan stock. West of central Nepal and up to Gilgit in northwestern Kashmir, the whole of the Lesser Himalayan belt is inhabited by the people of the "khas" stock, believed to have come in several waves from central Asia. In the eastern Himālaya (Arunachal) live people of the Tibeto–Burman stock who came from southeast Asia, primarily Myanmar. The Himalayan province is today a melting pot of peoples who have come from different parts of the subcontinent.

The people of the Himālaya have not only witnessed the tectonic development of their mountain in the present phase of its geological evolution but also borne the consequences of the quickened paces of the earth's processes, including the ever-threatening natural hazards. If the harsh conditions of the terrane and severity of climate determined their lifestyles, the inexorable grind of the earth's processes have made them both resilient and stoic. It will be interesting to know of the making of the mountain to which they belong. The story begins with the landmass of India breaking loose from the continent of Africa, drifting northwards and finally colliding with the continent of Asia. It was this collision of continents that gave birth to the Himālaya. However, it took several million years for the emergence and evolution of the highest and youngest mountain of the world.

(A)

(B)

(C)

**Figure 1.13** People of the Himalayan province. (A) The dwellers of the northern territory belong to the Tibetan stock of the Mongolian race; (B) A majority of the Lesser Himalayan people belong to the "khas" stock who came from central Asia. (C) In eastern Arunachal Pradesh, live people of Tibeto–Burman stock.

# 2

# *India Collides with Asia*

## India Breaks Away From Africa

There was a supercontinent called Gondwanaland in the southern hemisphere. India was an integral part of that ancient landmass. It split apart during the Mesozoic period (between 205 and 65 million years

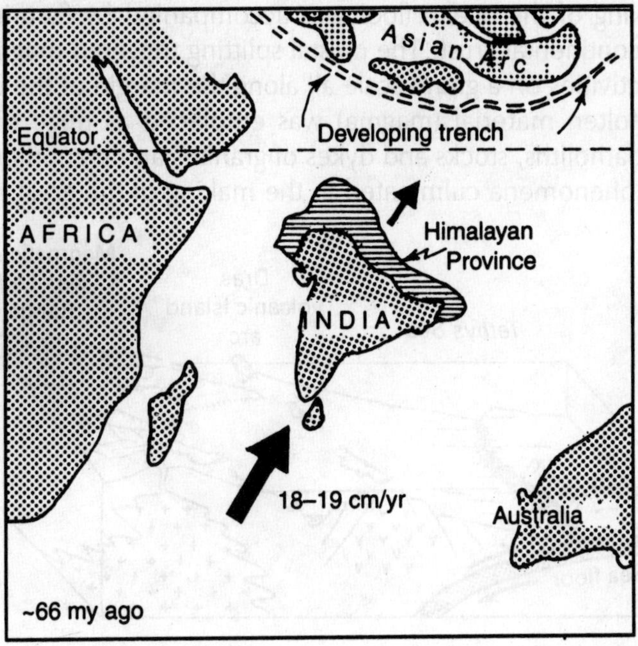


**Figure 2.1** India broke away 86–87 million years ago from Madagascar and moved northeast towards Eurasia. The leading edge of the Indian plate comprised a thick succession of sedimentary rocks interbedded with lavas. In front of Asia lay a chain of volcanic islands.

ago) to form what we now know as South America, Africa, Australia, Antarctica, Madagascar and India. A part of the landmass separated from the continent of Africa 100 to 105 million years ago and moved eastwards. Sometime between 87 and 86 million years ago, India broke away from Madagascar and drifted northeastward, and it converged at the rate of 18 to 19 cm per year towards the continent of Eurasia. It rode as a passive passenger on the crustal plate that formed the floor of what is known as the Tethys Ocean (Figure 2.1).

The leading edge or frontal part of the drifting Indian plate comprised of a prism-shaped succession of sedimentary rocks resting on the foundation of a complex of metamorphic and granitic rocks older than 2 500 million years. The sedimentary rocks and interbedded lavas belong to the ages earlier than about 1 600 to 65 million years BP.

## Convergence of India and Asia

As India approached Asia, its sediment-loaded front sank gradually. The part immediately to the south of Asia also sagged conspicuously, giving rise to an elongated, deep depression—an oceanic trench. The initiation of the sinking of the ocean floor was accompanied by fissuring of the adjoining continental crust. The crustal splitting triggered magmatic and volcanic activities on a grand scale all along the margin of the continent of Asia. Molten material (magma) was emplaced as huge discordant bodies of batholiths, stocks and dykes of granites and granodiorites. The magmatic phenomena culminated in the making of the great mountain

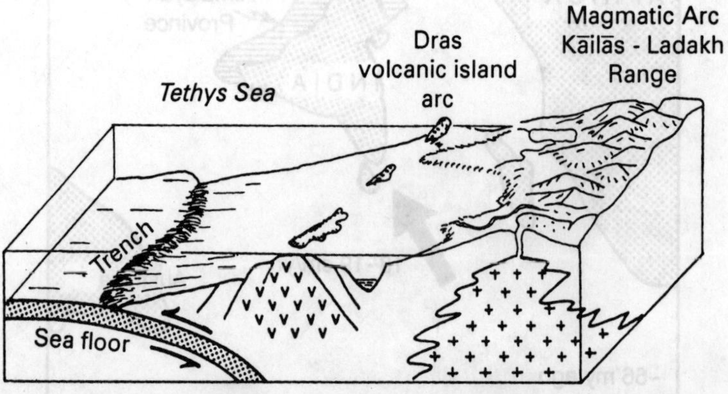


**Figure 2.2** The movement of India closer to Asia resulted in the sinking of the ocean floor and the formation of an oceanic trench just south of its southern margin. The fissuring of the crust triggered magmatic activities, resulting in emplacement of granite bodies. Simultaneously, lavas welled out through the fissures, giving rise to a chain of volcanic islands. The sunken floor, became an oceanic trench, and was rapidly filled with sediments.

arc stretching from Kohistan in northern Pakistan through Ladakh and Kailas to Gangdese in southern Tibet. The broadly contemporaneous volcanic activities are represented by lavas of andesite, basalt and rhyolite and explosion-generated fragmental material (agglomerate and ash). Lavas also erupted on the ocean floor of cold water, as borne out by the pillow-like structures they show and the conical seamounts that dot the deep ocean floor (Figure 2.2). As a matter of fact, a chain of volcanic islands developed, extending from Chalt in Kohistan through Astor and Dras in Ladakh to Shigatse in southern Tibet. The volcanic island chain possibly extended southeast in the Siang (Brahmaputra) Valley in eastern Arunachal Pradesh.

## Docking of India With Asia

Ploughing through the thick pile of sediments that had rapidly accumulated in the oceanic trench and on the ocean-floor, the drifting India touched and docked with the island arc in the Kohistan–Kargil sector (Figure 2.3). India had travelled more than 7 000 km since it had broken away from Africa 20 to 30 million years earlier. As the pressure of the converging continents grew, the sediments of the ocean-trench and the ocean-floor together with the volcanic rocks of the island arc were pushed onto the continental margin of India in Waziristan and Khurram in northern Pakistan and in the Cuafiang area in southern Tibet. The fossils contained in the sedimentary rocks indicate that this happened nearly 65 million years ago.

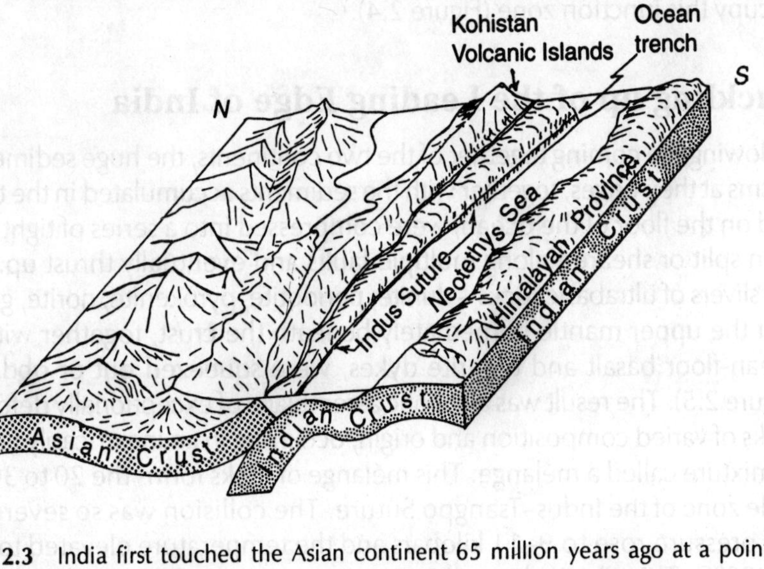


**Figure 2.3** India first touched the Asian continent 65 million years ago at a point in Kohistan in northwest Kashmir.

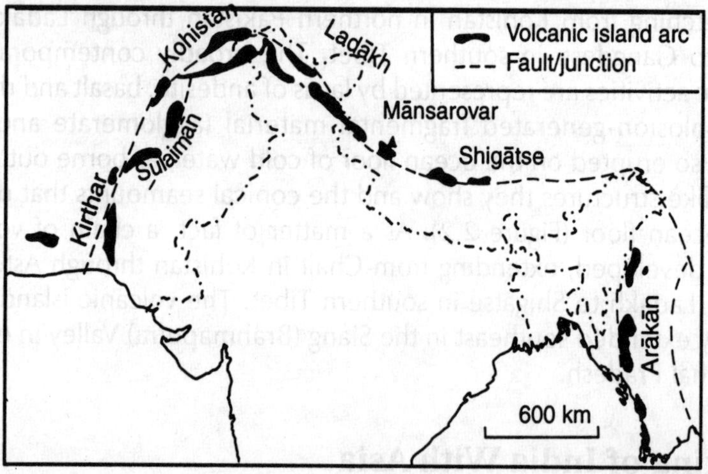


**Figure 2.4** The zone of collision between India and Asia, known as the Indus–Tsangpo Suture, is characterized by, among other things, a chain of volcanic islands that lie immediately to the south of the Asian continental margin.

The water of the now greatly shrunken ocean – the Tethys – was driven out, even as the rock assemblages of the ocean floor and the sediments of the ocean trench, together with the volcanic rocks of the island arc, were tightly compressed by the colliding continents. It took nearly 10 million years – from 65 to 55 million years – for the welding or amalgamation of the two continents. The collision zone, marking the Asia–India junction, is known as the Indus–Tsangpo Suture. The rivers Tsangpo and Sindhu today occupy this junction zone (Figure 2.4).

## Buckling up of the Leading Edge of India

Following the coming together of the two continents, the huge sedimentary prisms at their edges, together with the sediments accumulated in the trench and on the floor of the ocean, were compressed into a series of tight folds, then split or sheared along multiple faults and eventually thrust up. Even the slivers of ultrabasic rocks – dunite, peridotite, pyroxenite, norite, gabbro – of the upper mantle immediately beneath the crust, together with the ocean-floor basalt and dolerite dykes, were squeezed out or obducted (Figure 2.5). The result was a chaotic assemblage of exceptionally deformed rocks of varied composition and origin, occurring in an inextricably tangled admixture called a mélange. This mélange of rocks forms the 20 to 30 km-wide zone of the Indus–Tsangpo Suture. The collision was so severe that the pressure rose to 9–11 kilobars and the temperature elevated to 350° to 420°C. This physical condition is indicated by the presence of such minerals as glaucophane, jaedite, lawsonite, prehnite, etc., in the resulting blue-schist rocks.

The compression generated by the collision was so severe that the leading edge of the colliding Indian crust buckled up all along its length, giving rise to a domal upwarp. This is now discernible from Nimaling in Ladakh, through Raksastal in the Mansarovar region, to Lhagoi-Kangri in northeastern Nepal. Even the deeper part – the basement rocks comprising 475 to 550 million years old gneisses – were exhumed and placed against the younger rocks of the continental margin of Asia. It seems that the collision caused the bending down of the Indian lithosphere. (The lithosphere is the relatively rigid outer shell of the earth, which includes the continental and oceanic crusts and the upper part of the mantle lying above the softer interior, called the asthenosphere). Most of the earth scientists believe that the Indian plate plunged down and slid under the Asian plate for hundreds of kilometres beneath Tibet. The double-the-normal thickness of the Tibetan crust and the great elevation, (>5000 m) above sea level, of the Tibetan plateau are attributed to the under-thrusting of India below Tibet. In my perception, the buoyancy of the comparatively lighter rocks that make the northern frontal part of the colliding Indian continent would not have allowed the Indian plate to slide under Asia. The buckling of its front testifies to this fact. There was attendant detachment of the lower heavier part and buckling up of the comparatively buoyant upper crust. While the lower part is believed to have slid under Tibet for some distance, the upper part became an upwarp. Deep seismic reflection surveys carried out along the Nepal–to–Tibet line, however, do not show

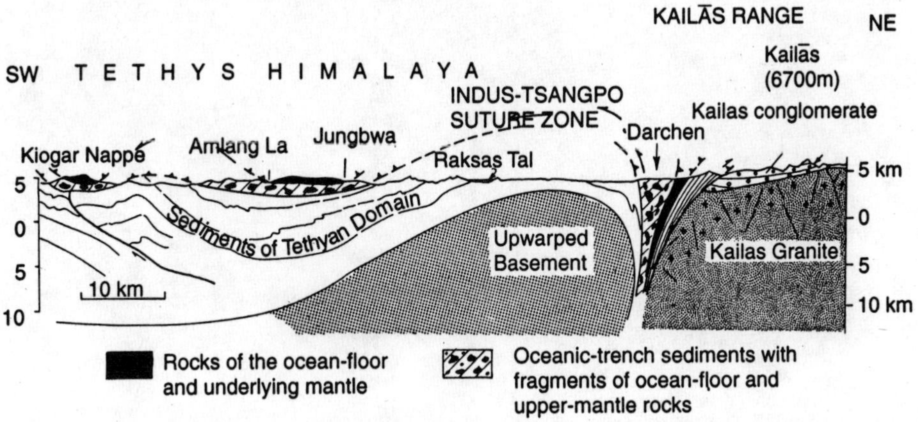


**Figure 2.5** The pressure of the India and Asia collision was so strong that the leading edge of the Indian plate buckled up into an elongated upwarp. The upwarp exhumed even the basement rocks in the anticlinal domes.

the Indian plate extending north beyond the Indus–Tsangpo Suture. On the contrary, the Indian crust is seen ramping downwards and forming an anticline with a duplex structure. In the Kohistan sector, where the collision was most severe and protracted, stacks of broken-up basement rocks popped up and were propelled southwards.

## Impact of Collision in Mainland India

The impact of the collision was also felt by the hind part of the Indian plate—in the Deccan domain of west-central India. A stupendous volume of basaltic lavas erupted in the short duration of one million years, 65 to 66 million years ago. The great pile of lavas then covered nearly a half-a-million square kilometre area of the Malwa and Deccan regions. The lavas erupted through fissures that had developed following the continental collision. Some earth scientists believe that the Deccan volcanism occurred when the northward-moving Indian plate crossed a rising column of buoyant hot lava, forming blazing fountains called the Reunion Hotspot.

The flood of hot lavas and the rains of hot volcanic ash and granules must have burnt down forests along with their wildlife, including the great dinosaurs that then reigned supreme in central India. It must have been a wholesale extermination of the heavy-footed, slow-moving animals and of the vegetation. The volcanoes must have poured into the atmosphere an enormous quantity of gases, including carbon dioxide and particulates, bringing about considerable climatic change at the turn of the Mesozoic era.

This, then, was the scenario of the time when the long process of the birth of the Himālaya commenced.

# 3

# *Water-Divide and Establishment of the Himalayan Drainage*

Following the India–Asia collision, the leading edge of the colliding Indian continent buckled up to form a low-amplitude upwarp all along its northern fringe (Figure 3.1). Simultaneously, the southern end of Asia

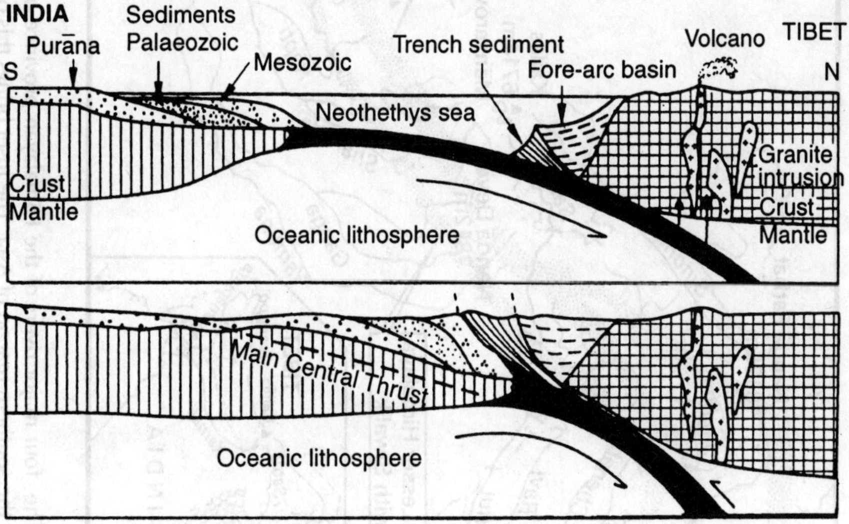


**Figure 3.1** Following the India–Asia collision, the sediments that filled the ocean-trench and covered the ocean-floor, along with the ocean-floor rocks, were squeezed into the zone of collision—the Indus–Tsangpo Suture.

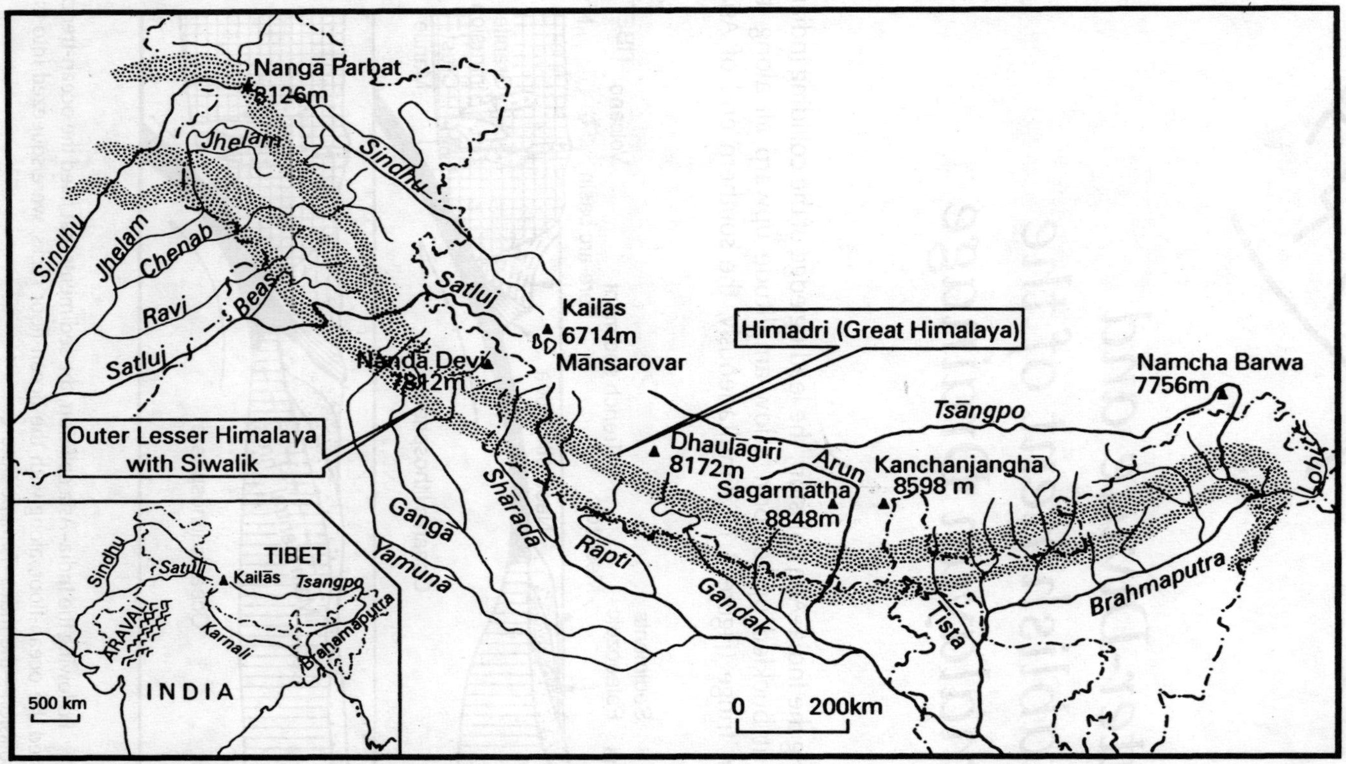


**Figure 3.2** The four major rivers of the Himalayan province originated from one area—the Kailas-Mansarovar tract in the water-divide that evolved along the southern margin of Asia. The inset brings this out clearly. The shading depicts the present-day high mountain ranges.

adjoining India was uplifted as a ridge. This zone of uplift embraced the terranes of Ladakh, Kailas and Gangdese. The Kailas-Mansarovar area formed a knot of sorts. The rivers that originated from this area flowed in different directions—the Sindhu went northwest, the Satluj flowed southwestwards, the Karnali (Ghaghara in the plains) took a southerly course, and the Tsangpo (Brahmaputra in Assam) went eastward (Figure 3.2)

The drainage pattern of the Himālaya was thus established soon after the amalgamation of the two continents. The belt immediately north of the collision zone formed the water-divide, but it was of no great elevation. The rivers that originated at the outset continued to flow in the channels they had carved out, despite the high barriers of mountain ranges that developed subsequently. Evidently, the rivers kept their channels open by cutting deeper and deeper as the mountain ranges rose higher and higher in the course of time.

## Resurgence of Tectonic Activity

The upwarping of the southern margin of Asia, leading to the development of a water-divide, is related to the crustal movement accompanying the intrusion of granite bodies all along the belt from Ladakh through Kailas to Gangdese, and in the adjoining areas of northern Nepal and Bhutan. This happened in the Middle Eocene epoch—36 to 45 million years ago. This, then, was the time when the great rivers of the Himālaya – the Sindhu, Satluj, Karnali, Arun, and Brahmaputra – were born in the highland that sloped southwards from the Ladakh–Kailas–Gangdese water-divide. The mountain, Himālaya, had not yet emerged.

## Sagging of Indian Crust

While the leading edge of the Indian crust bulged upwards, its hind part along what is now identified as the Outer Himālaya, sagged downwards to form a depression or foreland basin (Figure 3.3). This depression may be described as the *Sirmaur Foreland Basin*. The Sirmaur Foreland Basin was connected to the Indian Ocean through Sindh in the southwest and through Tripura–Meghalaya in the east. Likewise, another depression developed immediately beyond the crustal upwarp related to the zone of continental collision. This depression – the *Sindhu Basin* – was invaded by the water of the Neotethys Ocean. Perhaps a larger part of it represents the remnants of the sea that was forced to vacate when the two continents collided. The two depressions along the northern and southern margins of the Himalayan province developed presumably contemporaneously in the Middle Eocene epoch, 36 to 45 million years ago.

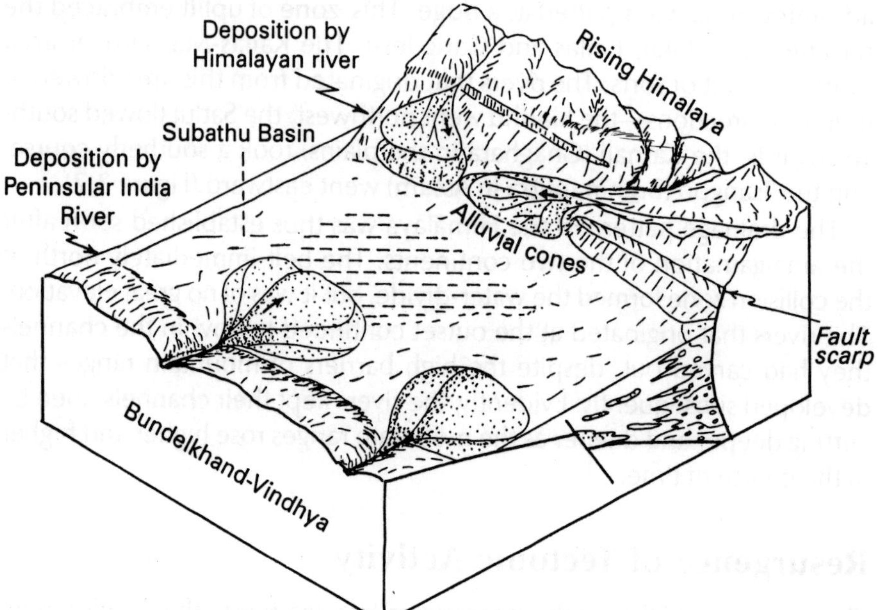


**Figure 3.3** Formation of the foreland basin due to the sagging of the Indian crust along the southern margin of the Himalayan province. The sagging was accompanied by faulting down of the basin floor and followed by the accumulation of sediments.

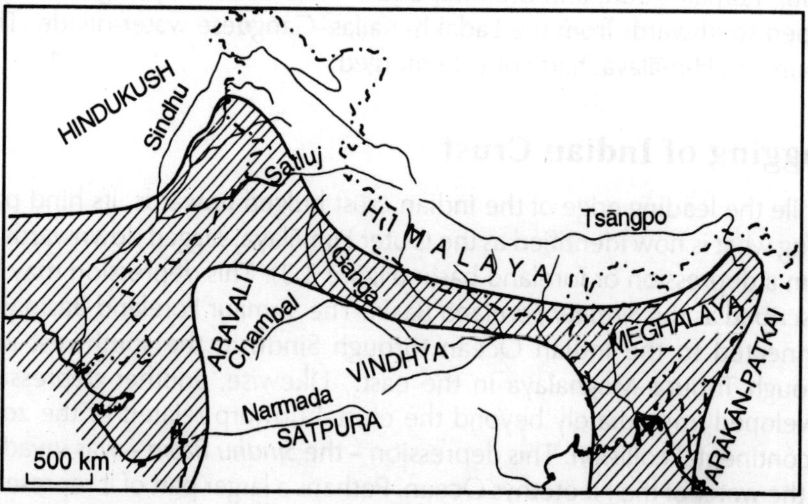


**Figure 3.4** The extent of the Sirmaur Foreland Basin along the southern margin of the Himalayan province; it was connected to the Indian Ocean. In the north, the Sindhu Basin occupying the collision zone partly represents the remnants of the Neotethys Ocean that was forced to vacate.

Sediments piled up rather rapidly in the Sirmaur basin, giving rise to the succession of olive-green and grey shale interbedded with siltstone, sandstone and shelly limestone. This assemblage of sedimentary rocks is known as the Subathu Formation (see Figure 6.4). This shallow basin covered parts of the Ranikot Hills in Sindh, Palana in Rajasthan, the Salt Range in Pakistan, the Jangalgali area in Jammu, Subathu in Himachal Pradesh, Bhainskoti in Nepal, Dalbuing in eastern Arunachal Pradesh, and the Jainti Hills in Meghalaya (Figure 3.4). In the relatively warm water, that was agitated by waves and currents, existed a variety of marine life (Figure 3.5), including algae, bryozoans, foraminifers (*Nummulites* and *Assilina*), diatoms, gastropods, pelecipods and fish. Their fossil remains occur in the sedimentary succession of the Subathu Formation and its equivalents.

## Immigration of Eurasian Animals

The docking of India with Asia formed land bridges across the two landmasses and opened the gates to the influx of a variety of terrestrial animals from different parts of Eurasia, including central Asia, Mongolia, China and Tibet. Suddenly, a large variety of vertebrate animals such as even-toed artiodactyls (goats, sheep, deer), turtles, crocodiles and fish made their dramatic appearance in great numbers in the Subathu domain of the foreland basin. This is evident from the fossils finds in the Kalakot area of Jammu, the Kuldana and Kirthar areas in Pakistan, and the Irrawaddy Valley in Myanmar. Significantly, in the Jammu– Kuldana part of the foreland basin, the sedimentary layers (of the Subathu Formation) containing mammalian fossils lie just above the layers characterized by fossils of marine molluscans. Clearly, soon after the coming together of the landmasses of India and Asia, the animals that lived in southern Tibet crossed over to the Indian landmass. These animals bear a striking resemblance to, and exhibit a close affinity with, those of Eurasia. Apparently, these animals must have had some connection through intermigration. In other words, the paths of immigration across the emerging Himālaya had been established as early as about 49 million years ago. Not only did the four-footed animals of the vertebrate clan arrive, but also the frogs which are allergic to salt-water. These frogs hopped their way across the Himalayan province and reached as far south as central India—where their remains are found buried in the lake sediments, sandwiched between the lava flows of the Deccan Volcanic Province.

## Evolution of Whales

One of the terrestrial mammals took to living permanently underwater in the sea, and eventually evolved to become a whale. This marine mammal

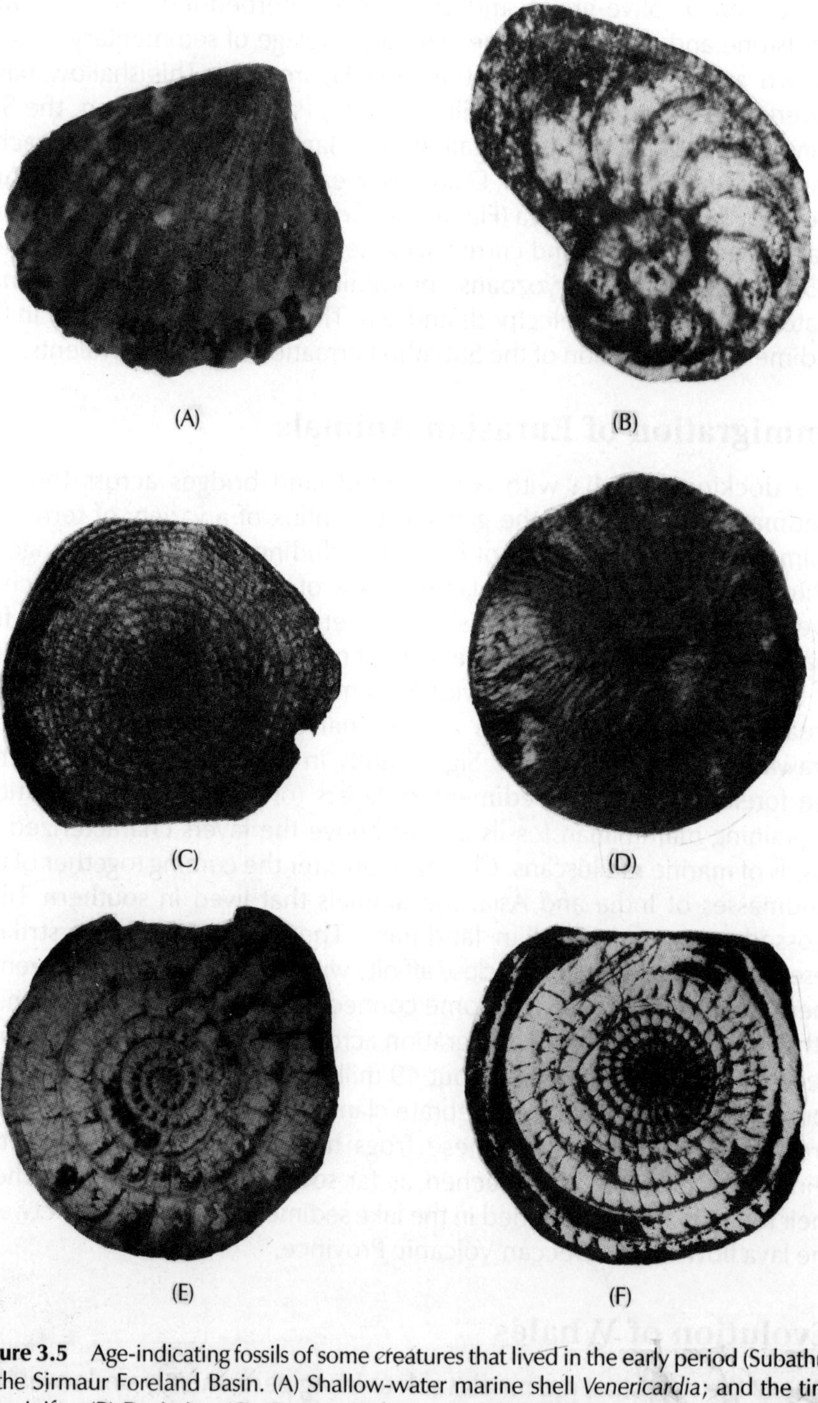


**Figure 3.5** Age-indicating fossils of some creatures that lived in the early period (Subathu) of the Sirmaur Foreland Basin. (A) Shallow-water marine shell *Venericardia*; and the tiny foraminifers (B) *Daviesina*, (C)–(D) *Nummulites*, and (E)–(F) *Assilina*. *(Photos: N S Mathur)*

**Figure 3.6** A terrestrial mammal that took to living permanently in the sea and became a marine whale, is *Ambulocetus*. *(Sketch by Sudip K Paul)*

is *Himalayacetus subathuensis*. Another such was *Ambulocetus*. Their remains were unearthed from the sedimentary succession of the Subathu Formation in the Jammu area (Figure 3.6). The latter creature represents the transition from terrestrial to marine mammal. This took place about 53 million years ago. Thus, the Simaur Foreland Basin of the Indian subcontinent is the cradle of the whales that are now found in oceans the world over.

## Formation of Deposits of Mineral Fuel

In the early stages of development of the Sirmaur Foreland Basin, vegetable debris derived from the forests covering the adjacent lands accumulated in the deeper parts along with fine-sized sediments. Increasing pressure and temperature, due to the growing pile of sediments converted the vegetable matter into coal through the intermediary stage of lignite. Deposits of lignite are seen at Palana in Rajasthan, of coal and lignite at Kalakot in Jammu, of coal in the Mikir Hills in Assam, and on the southern flank of the Meghalaya Hills.

In the Kangra District, in Himachal Pradesh, methane gas has been emanating out of the ground from time immemorial. The gas seeping out at Jwalamukhi is worshipped as an eternal divine flame (≡ *jwāla*). The petroleum gas must have been produced from organic matter in the quiet environment, deficient in or devoid of oxygen. There, the bacteria broke down the organic matter of plants and tiny marine animals (such as algae, fungi, diatoms, foraminifers, radiolarians, etc.) and, through a series of chemical changes at temperatures of 80 to 100°C, converted them into oil and gas. Certain clay minerals and metallic elements played the role of catalysts. Petroleum gas and oil and their equivalents, belonging primarily to the epoch Early to Middle Eocene 65 to 40 million years ago, were formed in the horizon of the Subathu Formation. Later, these fluids migrated upwards and were trapped in the younger sediments that succeeded the petroleum-producing horizon.

## Beginning of Continental Sedimentation

A strong tectonic upheaval overwhelmed the northern part of the Indian subcontinent. The sea was driven out from the Himalayan province forever. Marine sedimentation came to an end in the aftermath of this Middle Eocene tectonic movement, in both the Sindhu Basin at the northern margin and the Sirmaur basin at the southern margin of the Himalayan province. This happened about 42.6 million years ago in the Sirmaur Foreland Basin. After a long break of several million years, finer sediments and debris brought down by rivers and streams started accumulating in their channels and floodplains (Figure 3.7). This was the beginning of a new cycle of sedimentation on the continental basin by rivers that flowed in the south, southwest and westsouthwest directions in Nepal, in the southeastern and southwestern directions in Himachal Pradesh, and in the easterly direction in Pakistan. A new drainage system, in radical contrast to the previous one, was established. In place of the northerly drainage, the rivers now flowed southwards. Evidently, the Himalayan province had risen up as a high watershed and had started contributing the sediments generated by erosion and denudation to the rivers.

As the young mountain rose higher and higher, the pace of erosion became progressively more strident. Sediment-loaded rivers brought down

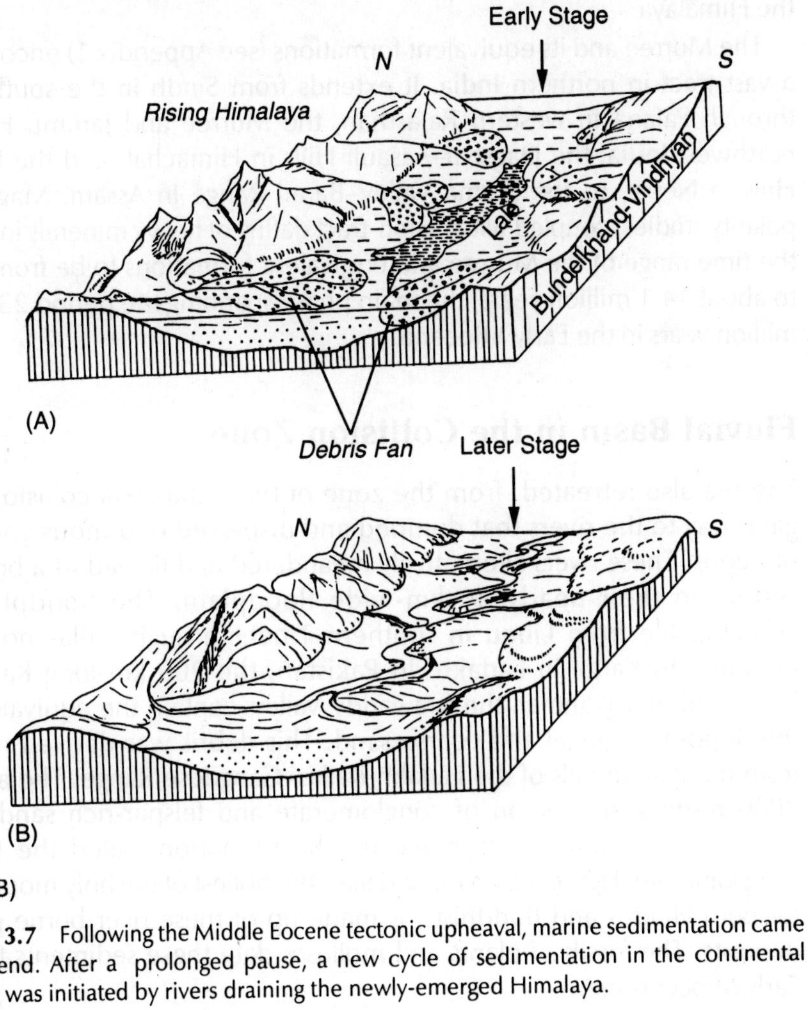


(B)

**Figure 3.7** Following the Middle Eocene tectonic upheaval, marine sedimentation came to an end. After a prolonged pause, a new cycle of sedimentation in the continental setting was initiated by rivers draining the newly-emerged Himalaya.

huge quantities of detritus and dispersed them on their floodplains that the Sirmaur Foreland Basin had by then become (Figure 3.7). The geologists describe this thick accumulation as the Murree and Dharmasala Formations. In the early stage, meandering rivers deposited purple, maroon and red clays and sands, and in the later stage preponderantly grey, muddy sand characterized by very tiny fragments of rocks. These rock fragments and heavy minerals like garnet and epidote (occurring in very small quantities) betray their parentage.They indicate that the sediments were derived from low-grade metamorphic rocks and associated hardened sedimentary rocks of the Himalayan province. This implies that in the Himalayan domain the lower-grade metamorphic rocks – which were formed at some depth – had been uplifted and exposed to denudation. In other words, higher ridges prone to erosion had emerged by this time in the Himālaya.

The Murree and its equivalent formations (see Appendix 1) encompass a vast tract in northern India. It extends from Sindh in the southwest, through Palana in western Rajasthan, the Murree and Jammu Hills in northwest India, the Dagshai–Kasauli Hills in Himachal, and the Dumri Hills in Nepal, to the Surma Valley–Barail Range in Assam. Magnetic-polarity studies, coupled with fission-tract dating of heavy minerals indicate the time range of the Murree and equivalent formations to be from 23.6 to about 14.1 million years, the main phase spanning the period 23 to 18 million years in the Early Miocene.

## Fluvial Basin in the Collision Zone

The sea also retreated from the zone of the India–Asia collision and gave way to the rivers that dumped and dispersed enormous volumes of debris. These rivers alternatively meandered and flowed as a braided system in their nearly 30 km-wide floodplain. The floodplain is recognizable from Liuqu in southern Tibet, through Kailas north of Kumaun, to Kargil in Ladakh. In Pakistan, the 200 km-long Katawas basin and, in Myanmar, the Irrawaddy Valley, contain the equivalent of the deposit of gravel and coarse sand. This debris was derived mostly from the granite hills of the Ladakh-Kailas–Gangdese Range. The almost 2000 m-thick succession of conglomerate and felspar-rich sandstone (arkose) with minor shale make up the formation called the Kailas Conglomerate (Figure 3.8). Mount Kailas, the holiest of the holy mountains for both Hindus and Buddhists, is made up of these river-borne gravel deposits. The fossils of plants and molluscs date these sediments to the Early Miocene age.

(A)

(B)

**Figure 3.8** (A) The holy mountain, Kailas, is made up of debris deposited by a river in its floodplains. (B) Closer view of the Kailas. *(Photos: Anup Sah)*

(A)

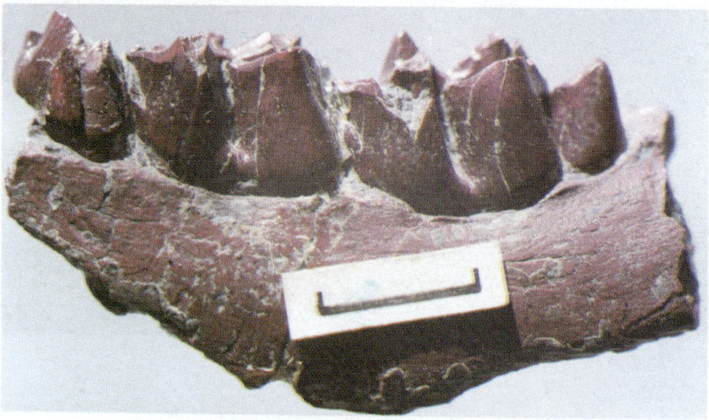

(B)

(C)

**Figure 3.9** Life in the floodplains of the Early Miocene Sindhu Basin in Ladakh. The plant fossil indicates a warm climate and a ground height of less than 1 700 m above sea level. (A) *Iberomeryx,* an even-toed animal (deer or goat), (B) *Lophiomeryx* tooth of another even-toed animal, (C) *Trachycarpus* palm. *(Photos: A & B: A C Nanda; C: A K Sinha)*(A)(B)(C)

## Life and Climate Conditions

The impressions of palm leaves and fossils of rosewood and charophytes in the shale of the Kailas Conglomerate in Ladakh (Figure 3.9) indicate the prevalence of a warm–moist climate in the Sindhu–Tsangpo valleys in the Early Miocene time. The presence of caliche in the sediments in the lower horizons of the Murree Formation in Himachal Pradesh indicate that the climate then was semi-arid. In other words, the slopes of the low hills of the Himalayan province in the Early Miocene time were covered by palm trees and associated vegetation. At present the palms occur commonly below the 1 700 m elevation. This implies that the Himālaya mountain, in the period of 23 to 16 million years ago, could not have been higher than 1 700 m above sea level.

In the warm–wet terranes in the valley of the Sindhu in Ladakh, lived fish, turtles, pythons, snakes, crocodiles, small-sized goats, deer, and rodents. Vertebrate animals, very similar to these, also inhabited the terrane of the Bugti Hills in southern Pakistan. It appears that these animals could and did freely migrate across the Himalayan province from Ladakh to Baluchistan or vice versa. This implies that the topography of the Himālaya province then was fairly gentle, and the relief mild enough to allow the migration of animals from one part to another. In the marshy woods of the Surma Valley, the Barail Hills in Assam, and the Irrawaddy Valley in Myanmar, lived the ancestors of the rhinos that we see today.

# 4

# *Emergence of the Himādri as a Mountain*

## Breaking of the Crust in the Himalayan Province

The Himalayan province was again overtaken by a severe convulsion of crustal movement in the Early Miocene epoch. The Indian crust ruptured along deep faults which were more than 2 400 km length. Of the several faults that developed, two are of tremendous import, as they played a very crucial role in the evolution of the Himādri or Great Himālaya. The rather gently-inclined fault, known as the Main Central Thrust (MCT), pushed up a great succession of rocks including the basement or foundation rocks that lay at the bottom. The thrust-up rock piles moved southwards, trampling over the sedimentary rocks of the Lesser Himālaya. In the north, a series of steep faults detached or decoupled the sedimentary rock pile of the Tethyan terrane from its hard foundation, the basement complex (Figures 4.1 and 4.2). This fault is known as the Trans-Himādri Fault (T-HF) in the central sector, the South Tibet Detachment Thrust in Nepal, and the Zanskar Shear Zone in Himachal. The MCT and T-HF seem to have evolved almost contemporaneously.

Between the MCT and T-HF the stupendous crustal block made up of high-grade metamorphic and granitic rocks was uplifted, giving rise to what eventually became the Himādri or Great Himālaya. In the northwestern sector, embracing the Chamba and Kashmir regions, the displacement along the T-HF brought southwards up to the border of the Himalayan province not only the Tethyan sedimentary pile but also the upper part of the basement (made up of low-grade metamorphic rocks). This uprooted basement slab forms the lofty Dhauladhar–PirPanjal Ranges overlooking the Śiwālik.

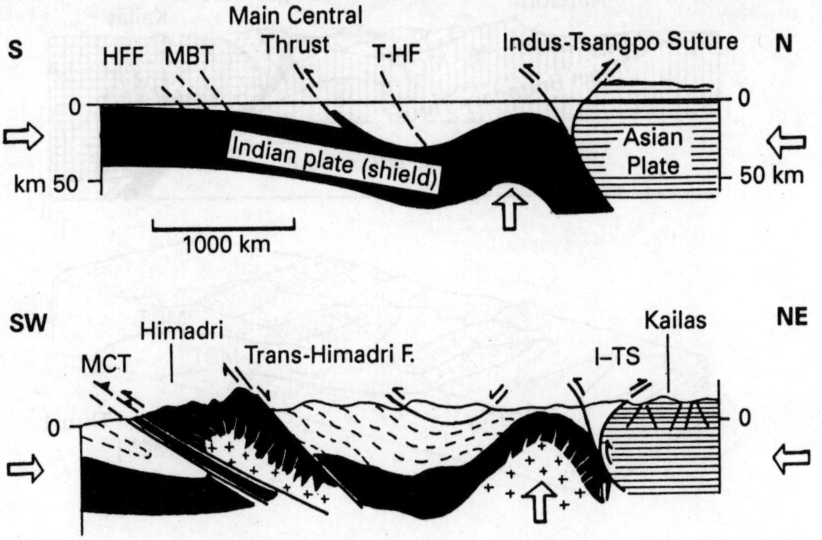


**Figure 4.1** In the Early Miocene period, the Indian crust — which had earlier buckled up at its frontal part — broke up along a series of faults. While the basement complex was thrust up along the Main Central Thrust (MCT), the sedimentary cover of the Tethyan domain was detached from the foundation and slipped northwards along the Trans-Himādri Fault (T-HF), which is known as the South Tibet Detachment Thrust in Nepal. In the far south, yet another crustal rupture was developing, which later became the Main Boundary Thrust (MBT).

## Making of the Himādri

As already pointed out, the foundation – the basement complex making the high mountain that later emerged as the Himādri – is made up of high-grade metamorphic rocks intruded by 540 to 470 million year-old porphyritic granites. The geologists describe this rock assemblage as the Vaikrita Group. Repeated and strong movements along the MCT gave rise to a wide zone of very strong shearing and pronounced milling of rocks, and the attendant development of an extremely dense texture (mylonitization). These rocks are characterized by a multiplicity of planes of gliding and sliding. The much-deformed and mylonitized rocks occur as overlapping sheets or slabs, in the manner of the imbricated tiles of a roof.

The tectonic movements were accompanied by high-grade metamorphism. Mineral assemblages of various rock types of the Himādri terrane indicate the pressure rising to 6 to 10 kb (locally even upto 12 kb) and the temperature in the range of 600° to 800°C. This high pressure and temperature must have developed at a depth of 25 to 35 km below the

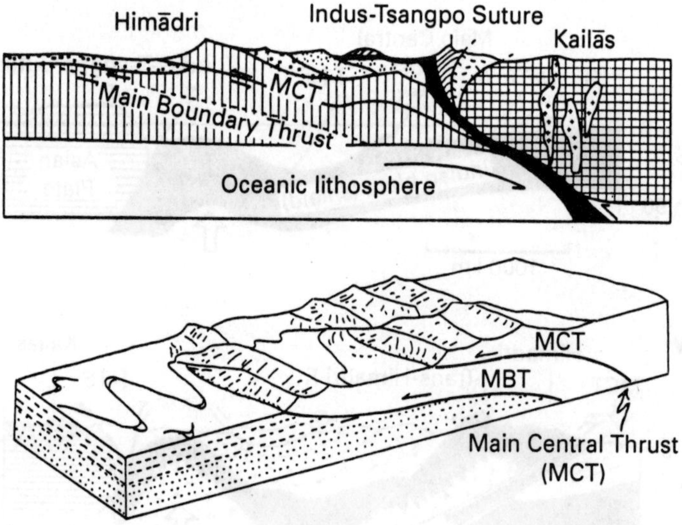


**Figure 4.2** (A) The diagram portrays the structural design and position of the Himādri in the Himalayan province. The basement was thrust up and the sedimentary succession, originally resting on the basement, slipped northwards. The MBT is beginning to develop in the south. (B) A block diagram showing the above development is after T Tokuoka and others, 1994. Following the India–Asia collision, the sediments that filled the ocean-trench and covered the ocean-floor, along with the ocean-floor rocks, were squeezed into the zone of collision—the Indus–Tsangpo Suture.

surface. This metamorphism occurred before the rocks were thrust up to form high mountains. Under the high temperature–pressure conditions at great depths, the rocks melted partially or differentially, giving rise to molten material of granitic composition. This phenomenon is called anatexis. Strikingly light in colour and characterized by black tourmaline, these granites are remarkable because of the presence of such minerals as sillimanite, garnet, cordierite, etc., which are characteristic minerals of the high-grade metamorphic rocks of the Vaikrita Group. These minerals thus demonstrate the formation of granites from the melting of high-grade metamorphic rocks. The molten granitic material penetrated very intimately, leaf-by-leaf, the gneisses and schists of the Himādri succession, giving rise to a mixed rock called migmatite. Formed as they were in the upper part of the Vaikrita succession of the Himādri, the granites intrude even into the overlying sedimentary rocks of the Tethyan domain. They occur as batholiths, stocks, laccoliths and dykes and locally form networks of veins. The holy peaks of Kedarnath and Badarinath in Uttaranchal are made up of the young granite that was intruded in the period 25 to 18 million years ago. The peak activity of granite emplacement took place about 20.5 to 21.5 million years ago. This age is indicated by fission-track dating of zircon and apatite, Ar/Ar isotope age of hornblende, U-Pb age of zircon, and Th-Pb age of monazite that occur in these granites.

The intrusion of the hot, mobile granite into the high-grade metamorphic rock was attended by a flux of heat and chemically active fluids in the contact zone. This phenomenon raised the grade of the already high-grade metamorphic rock in the upper part of the Vaikrita succession. The result was the unusual spectacle of high-grade metamorphic rocks occupying the top level of the succession. Under normal conditions, it should have been the other way round.

## Timing of the Emergence of the Himādri

As mentioned earlier, the granites were formed in the period 21 to 25 million years ago, mostly between 22 and 23 million years. The origin of these granites is intimately related to the severe deformation, including thrusting and sinking to great depth, of the rocks. The detachment of the Tethyan sedimentary pile from its foundation along the T-HF was accompanied by extension of the crust in the east-west direction. And this extension took place in the period 19.5 to 23.5 million years, or roughly 22 million years ago in the Early Miocene epoch. It is obvious that the Himādri emerged as a high mountain 21 to 22 million years ago.

## Southward Thrusting and Squeezing of the Basement Rocks

As the pressure of continental compression grew, the tectonic tumult in the Himalayan province rose to a climax. Quite a large part of the squeezed-up rock assemblage of the basement was pushed or thrust southward. The uprooted masses were displaced tens of kilometres away onto the Lesser Himalayan terrane of the sedimentary rocks, more than 1600 million years to about 540 million years old. They advanced as much as 95 to 140 km from their roots at the base of the Himādri. The uprooted rock masses comprising of lower-grade metamorphic rocks associated with granites of two generations (1800 to 2000 million years and 500 to 540 million years), occur as vast sheets called nappes or as smaller patches known as klippen. They partially cover the sedimentary pile of the Lesser Himalayan terrane (Figure 4.3). Later tectonic compression folded the overthrust sheets of these crystalline rocks concordantly with the underlying sedimentary succession into a series of folds that form the mountain ranges of the Lesser

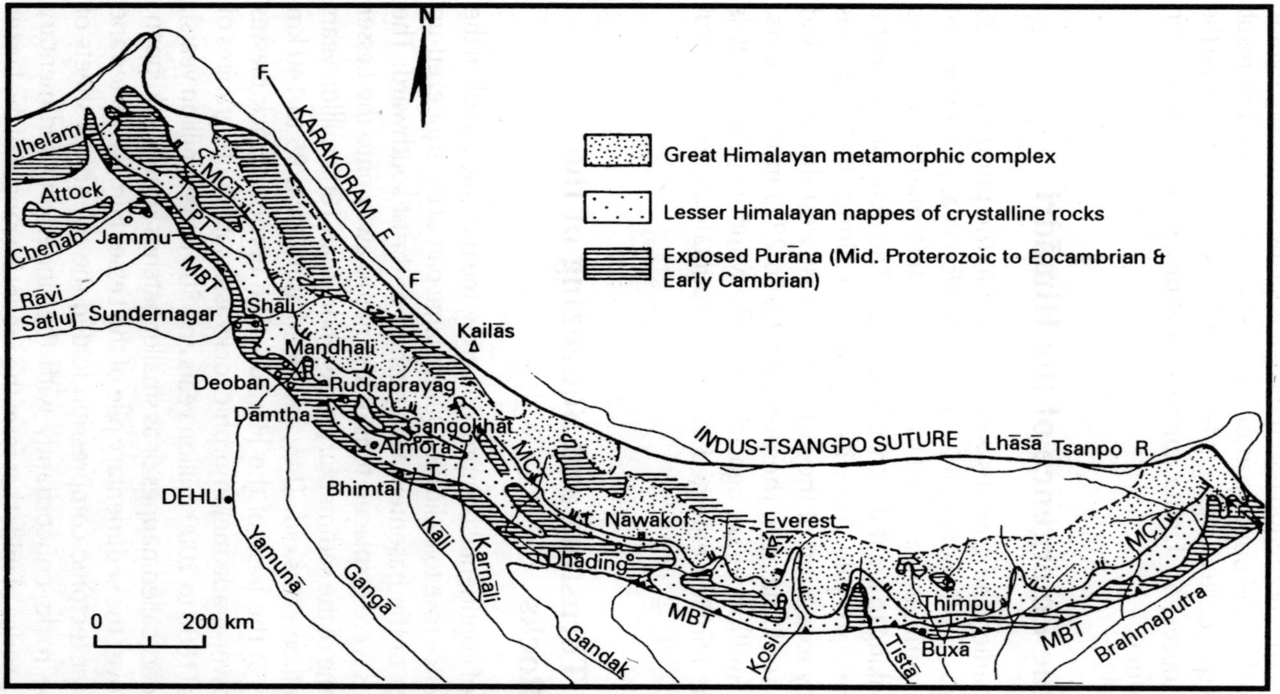


MCT - Main Central Thrust, MBT - Main Boundary Thrust, PT - Panjal T.

**Figure 4.3** The high-grade metamorphic rocks associated with 500–540 million years old porphyritic and 19 to 21 million years old light-coloured granites make up the Himādri (Great Himālaya) terrane. The lower-grade metamorphic rocks associated with 1800–2000 million years old porphyry and porphyritic granite and 500–540 million years granite form the Lesser Himalayan nappes and klippen. The nappes and klippen rest upon, and are concordantly folded with; the underlying Proterozoic-to-Early Cambrian sedimentary succession of the Lesser Himālaya.

Himālaya. On the synclinal mountains of this kind are located such townships as Shimla, Ranikhet, Almora and Darjiling.

The metamorphic rocks of the Lesser Himālaya occurring as nappes had a different pressure-temperature history—they were formed under conditions very different from those of the Himādri domain. This is but natural, for the two were formed in locations physically quite separated.

## Gravity Gliding in Northern Border

The anticlinal crustal upwarp, adjacent to the zone of collision of Asia and India, was also compressed and elevated during the Early Miocene movements. The sudden uplift of the already high ridge triggered gravity-induced gliding and sliding of the pile of ophiolites and ophiolitic mélange on the steep slopes. These ophiolites had earlier been squeezed out of the ocean floor and the oceanic trench when the continents had collided. They now lie helter-skelter upon and amidst the rocks of an altogether different environment. Spectacular blocks of the alien rocks in chaotic condition are seen in Malla Johar in northern Uttaranchal, at Spongtang in Ladakh, and a few places in Kohistan in Pakistan (Figure 2.5). The Tethys sedimentary rocks were severely compressed into tight to overturned folds, some of which were split by faults along their axial planes.

In brief, the orographic feature called Himādri or Great Himālaya first emerged in the Early Miocene time 21–22 million years ago. Both the Tethys realm to the north and the Lesser Himālaya terrane to its south experienced unprecedented tectonic turmoil. This is reflected in the extent of deformation and the resultant erosion of the uplifted mountain.

In this account of the formation of the suture zone between the colliding continents, the evolution of the Great Himālaya and the development of the Lesser Himālaya, frequent references have been made to old rocks belonging to different geological ages and formed in varied environments. It would be pertinent – and quite instructive – to take a quick look into the past to know what had happened in the cradle of the Himālaya before it emerged. In Chapter 5 we will explore the past in order to understand the embryonic development of the *"Nagadhiraj"* Himālaya.

# 5

# A Flashback into the Past

## Northern Continental Margin

It was the great pile of sedimentary rocks resting on the foundation made up of metamorphic and igneous rocks, along the northern continental margin of India, that evolved into the Himālaya as a result of severe compression, folding and splitting, following the collision of India with Asia about 65 million years ago. In this great sedimentary pile is written the long account of the events preceding the birth of the Himālaya (Figure 5.1).

It would be both interesting and instructive to read this account of the emplacement of sediments on the continental margin of India. It may be mentioned here that it was not an uninterrupted phenomenon. There

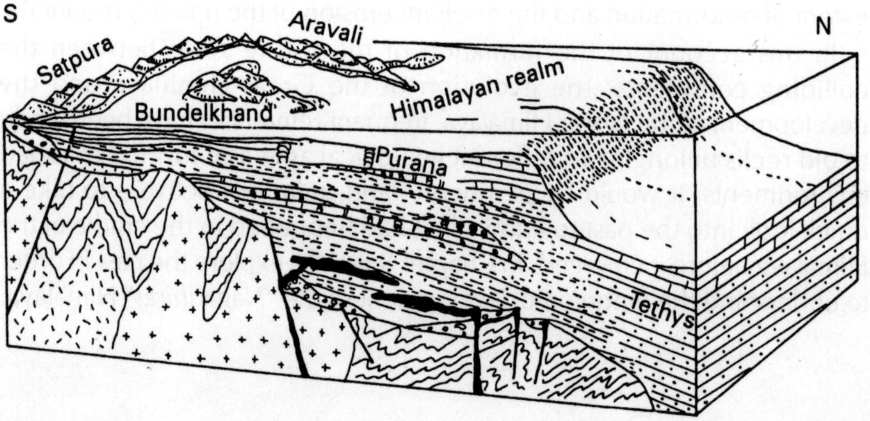


**Figure 5.1** This sketch portrays the nature of the pile of rocks of the *Purana Sea* that embraced the northern continental shelf of India. The succession of sedimentary rocks span the time from about 2000 million years to about 525 million years ago.

were many breaks or hiatuses that occurred at different times in different sectors. Each of these breaks implies a spurt of tectonic movement, resulting in the displacement, dislocation and deformation of rock masses and the concomittant speeding up of erosion in some parts and interruption or even cessation of sedimentation in others. To put it differently, before the great mountain emerged, the Himalayan province went through several convulsions of tectonic tumult, witnessed many episodes of fiery floods of lavas, and was injected many a time by hot molten material of granitic composition, even as the deposition of sediments in varied environmental settings continued sporadically. Life appeared quite early in the history of the Indian continent, but proliferated only after the beginning of the Cambrian period 570 million years ago. It is not easy to decipher the history of the Himalayan orogenic province owing to its bewildering tectonic complexity, a result of repeated and severe deformation and dislocation of rocks.

## Beginning of Sedimentation Under Disturbed Conditions

There was once a big sea sweeping the northern margin of the Indian continent and washing the foothills of the then high Aravali and Satpura mountains (Figure 5.2). Named the *Purana Sea* (Purāna = पुराण ancient), it

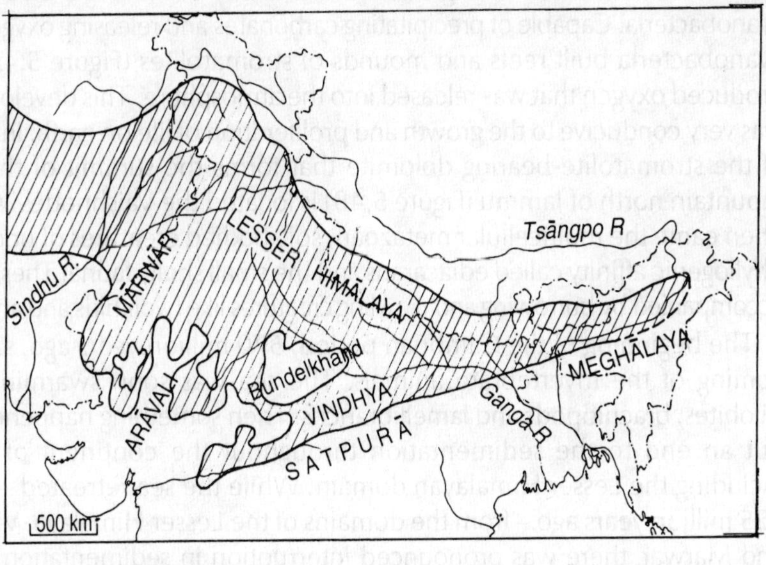


**Figure 5.2** A conceptual map showing the extent of the northern continental margin of India. It represents the spread of the sediments deposited in the Purana Sea during the Proterozoic-to-Early Cambrian temporal span.

had come into existence sometime around 2000 million years ago and existed until about 525 million years ago. The sea embraced the domains of the Himālaya in the north, the Vindhya in central India and Marwar in western Rajasthan. The floor of this shallow sea sloped northwards and westwards. The foundation (floor) was made up of nearly 2500 million year-old granite and gneiss and associated metamorphic rocks in the southern part, and of nearly 2000 million year-old porphyritic granite and porphyry in the northern (Himalayan) part. In the shallow basin of the Purana Sea were deposited sediments washed down from the highlands of the Peninsular Indian Shield by rivers that flowed in the northerly and northwesterly directions.

It was not a quiet beginning of the Purana sedimentation. There were interludes of tectonic disturbance that triggered underwater mass-movement including landslides, debris avalanches and turbidity currents in the particularly disturbed areas. These events are represented by diamictite or muddy conglomerate occurring amidst greywacke (muddy sandstone) and shale (Figure 5.3). Many parts of the sea basin witnessed volcanic activities—the lava welling out through fissures and faults in the ocean floor, and the volcanic ash and fragmental material blanketing the marine sediments.

Towards the middle of the Purana era, when tectonic tranquility prevailed and there was stability all around, carbonates of calcium and magnesium were precipitated, giving rise to limestone and local magnesite (Figure 5.3). The first life-forms were single-celled and many-celled organisms called cyanobacteria. Capable of precipitating carbonates and releasing oxygen, the cyanobacteria built reefs and mounds of stromatolites (Figure 5.4A) and produced oxygen that was released into the atmosphere. This development was very conducive to the growth and proliferation of life on earth. In a cave of the stromatolite-bearing dolomite that forms the summit of the high mountain north of Jammu (Figure 5.4B) is located the Vaishnodevi shrine. Then came the multicellular metazoan, soft-bodied creatures of uncertain phyllogenic affinity called ediacarans and the small shelly fauna. These were accompanied by burrowing and crawling creatures like hyolithids and annelids.

The beginning of the Cambrian period, 570 million years ago, saw the coming of the invertebrate animals. The sea was soon swarming with trilobites, brachiopods and lamellibranchs. Then something happened that put an end to the sedimentation throughout the continent of India, including the Lesser Himalayan domain. While the sea retreated – about 525 million years ago – from the domains of the Lesser Himālaya, Vindhya and Marwar, there was pronounced interruption in sedimentation in the northern (Tethyan) part of the Himalayan province. Before the sea vanished from the Lesser Himālaya-Marwar-Vindhya part, there was dessication in some tidal flats and embayments of the retreating sea. This is evidenced by

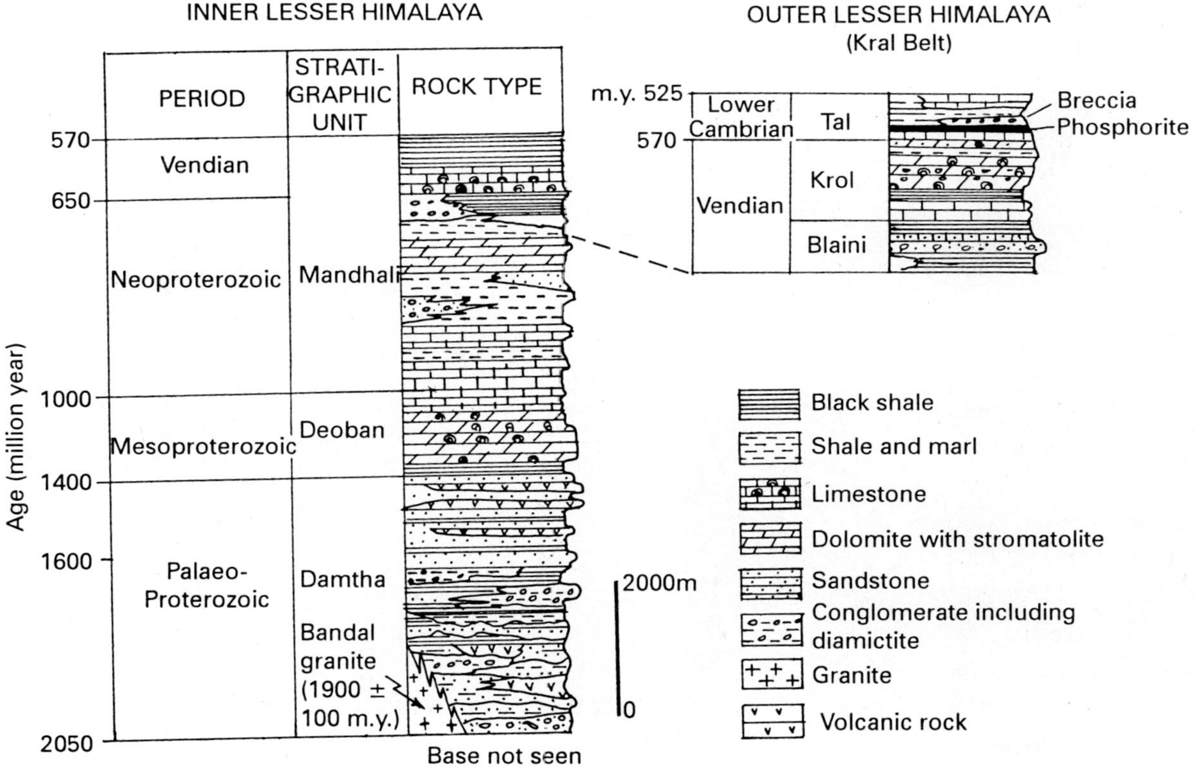


**Figure 5.3** The succession of sedimentary rocks that record the history of events in the Purana Sea in the Lesser Himalayan domain. The geologists subdivide the successions into lithological groups and formations and give specific names to the various units.

(A)

(B)

**Figure 5.4** (A) Cyanobacteria formed the stromatolites by precipitating and binding together carbonates and freeing oxygen. The latter considerably changed the nature of the atmosphere. (B) In the cave of the stromatolite-bearing dolomite north of Jammu is located the holy shrine of Vaishnodevi

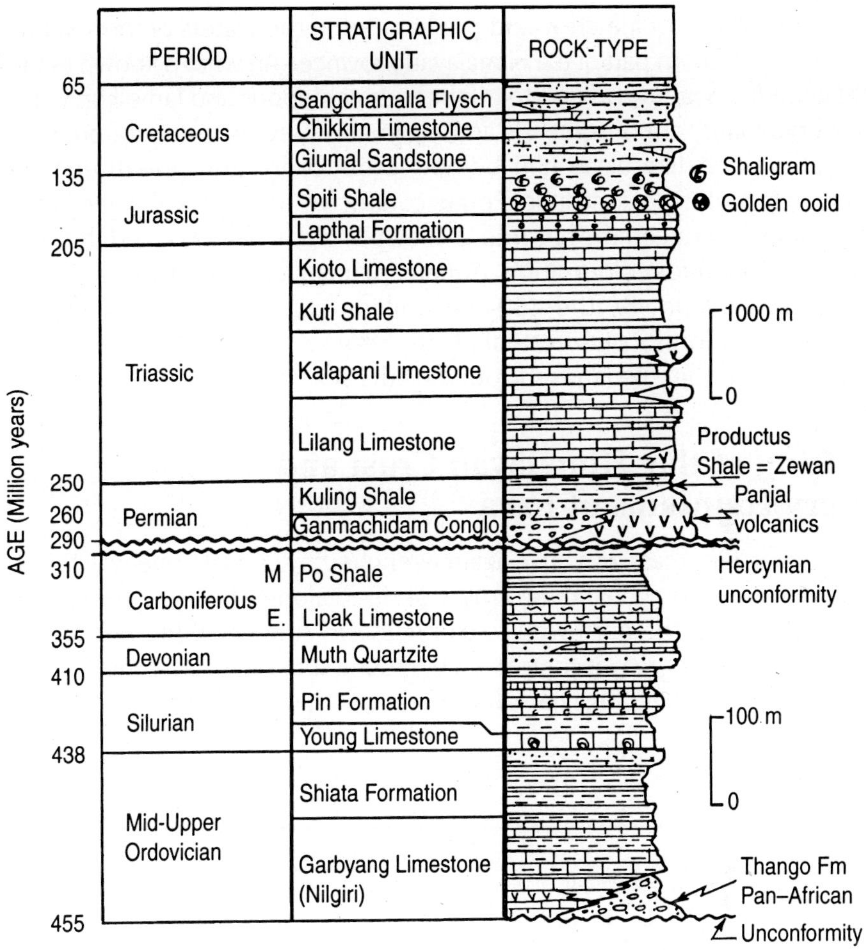


**Figure 5.5** The succession of sedimentary rocks in the Tethys domain in the northern Himalayan province, spanning the temporal period from the Late Proterozoic to the Late Cretaceous.

the deposits of salt and gypsum in Mandi in Himachal Pradesh, the Salt Range in Pakistan, and Nagaur in Marwar. These salt deposits are being mined today.

The cessation of sedimentation throughout the Lesser Himālaya and Peninsular India was roughly contemporaneous with the large-scale invasion of molten material of granitic composition 525 to 470 million years ago. Some parts of the Himalayan province witnessed volcanic activities also—during or soon after the events of granite magmatism. Evidently, the entire Indian continent had been overtaken by the tectonic upheaval that forced the Purana Sea out from the Lesser Himālaya, Marwar, Vindhya and many other parts of the Peninsular India. However, as already stated, in the northern (Tethyan) part, the sedimentation continued after interruptions of rather short though variable duration (Figure 5.5).

A rich variety of life grew and proliferated in the waters of the shallow sea in the northern part of the Himalayan province—in what is known as the *Palaeotethys Sea*. In addition to trilobites, brachiopods and lamellibranchs, the bryozoans, crinoids, corals and algae also grew in great abundance. Aquatic plants like algae had by now produced (rather freed) enough oxygen to make the atmosphere quite conducive to life. The oxygen in the atmosphere reduced the hazardous effects of ultraviolet rays, and the land was now open for colonization. Aquatic plants were the first to secure a foothold on land. By the 450–440 million-year period, the land was clothed with forests of terrestrial plants. Very soon, animals followed the plants to conquer the lands surrounding the sea.

## Rifting of the Himalayan Crust and Fiery Fountains of Lava

The Indian landmass was once again wrenched by crustal upheaval 310 to 280 million years ago, and the crust was rifted along a belt stretching from the PirPanjal in the northwest to southeastern Arunachal Pradesh in the east. The rupturing was accompanied by ground subsidence all along the belt. Seawater rushed into this narrow depression (Figure 5.6). Lava welled out

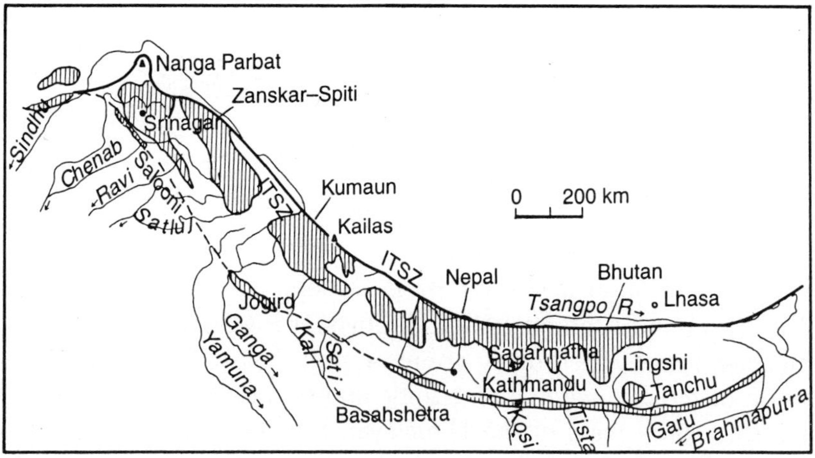


**Figure 5.6** Present-day extent of the sediments deposited in the Palaeotethys during the Palaeozoic era spanning the period from about 525 to nearly 260 million years ago. Rifting and attendant subsidence gave rise to a narrow depression in the south. Into this depression rushed the seawater of the Gondwanic world.

through fissures of the ruptured crust on a grand scale. Many explosive volcanoes hurled fragmental material around their fiery fountains, and the sediments accumulating in the narrow depression were covered by blankets of volcanic ash. Glaciers covering the highlands of the Peninsular Indian Shield during that period of the geological history of the Gondwanaland dumped their gravelly detritus into this seaway. There was thus an intermingling and interfingering of marine, volcanic and glacial material in the Early Permian time 290 to 270 million years ago all through the belt that today occupies the southern border of the Lesser Himālaya (Figure 5.6).

The floodplains of the northward-flowing rivers were then covered with dense forests of *Glossopteris* and *Gangamopteris* assemblages. Buried in the sediments of the floodplains and in the stream channels, these and other associated plants and their debris were eventually converted into valuable deposits of coal. Such deposits occur not only in the basins of the Godavari–Pranhita, Narmada–Son, Mahanadi and Damodar in the Peninsular Indian Shield, but also in the Eastern Himālaya east of the Arun River.

## Breaking Away of Tibet

The fissuring of the crust, with the attendant rampant volcanism culminated in the breaking away of the northern part of the Indian continent (Figure 5.7). This happened about 260–250 million years ago. The breakaway part comprised the microcontinents now known as Tibet, Iran and Turkey. The sea that opened up between India and this string of microcontinents is called the *Neotethys*. Sedimentation continued in the Himalayan part of the Neotethys—vigorously in some parts and haltingly

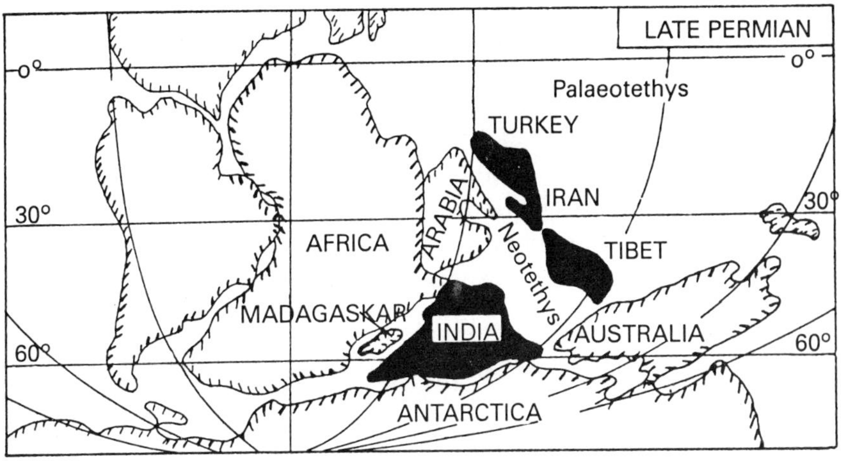


**Figure 5.7** The opening of the Neotethys Sea between India and the string of microcontinents represented by what are now Tibet, Iran and Turkey. The Tibetan microcontinent broke away from India 260 to 250 million years ago.

in others. Volcanic activity persisted in some sectors, such as the Kashmir–Zanskar region in the northwest and northern Nepal in the middle.

The Neotethys Sea was peopled by a great variety of animals including the cephalopods. The ammonites together with lamellibranchs suddenly acquired a pre-eminent position, marginalizing the thus far dominant brachiopods. The Triassic period 250 to 205 million years ago was the time of cosmopolitan faunal life, characterized by great diversity and prolificity. The ammonites formed very characteristic horizons of limestone and dolomite throughout the Himālaya. It is in the upper part of the succession of the ammonite-bearing limestone in the Sonmarg-Zanskar Range in Kashmir that the holy cave of Amarnath is located (Figure 5.8 A and B).

## Sinking of the Continental Edge

Then in the Jurassic period, 205 to 135 million years ago, the northern edge of the Indian continent started subsiding. The Neotethys Sea expanded and a large part of the continent including Saurashtra, Kachchh and Marwar in the west and southern Meghalaya in the east, came under the sway of the seawater. Sediments, plant remains and debris were washed down from the highland and deposited along with the marine sediments. One of the most conspicuous features of the sediments of this period is the very striking golden and reddish, spherical or ovoidal oolites and pisolites in the iron-rich limestones seen practically throughout the vast sea. Resembling fish roes in appearance, these golden-coloured grains are characterized by concentric shells of calcium carbonates. Another remarkable feature of immense interest is the concretionary ball or nodule formed around the nucleus of an ammonite fossil. Known as *'Shaligram'*, the concretionary ball (Figure 5.8C) is worshipped as a deity.

The Neotethys Sea continued to deepen along the northern periphery of India as it sank progressively (Figure 5.9). Into the deepened sea were deposited siliceous (radiolarian) and calcareous (foraminiferal) oozes and fine pelagic clays. They formed beds of chert, limestone and shale, respectively. Phosphatic nodules and glauconite mineral occurring in some of these rocks indicate that in the deep sea, the water was quite cold and deficient in oxygen.

## Precursors of the Impending Orogenic Revolution

While the frontal edge of the Indian plate subsided, its back portion in eastern India was rent apart and volcanoes spewed out lava and ejected rock fragments and ash, 117 to 115 million years ago. The Rajmahal Hills, formed in this manner, was soon covered by Gondwanic vegetation. The northward-

(A)

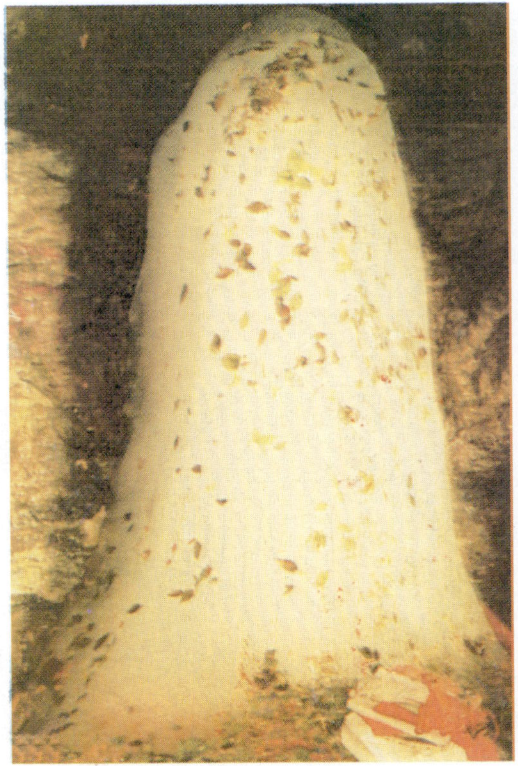

(B)

(C)

**Figure 5.8** (A) The Amarnath Cave in the ammonite-bearing (230–205 million years old) Upper Triassic limestone in the Zanskar–Sonmarg Range. (B) The exquisite ice stalagmite in the cave is worshipped as a deity *(Photos: B L Dhar)*. (C) *Shaligram*, a peculiarity of the 205–135 million year-old Jurassic sediments in the northern Tethyan Zone, is a concretionary ball or nodule having an ammonite fossil as its nucleus. *(Photo: A K Sinha)*

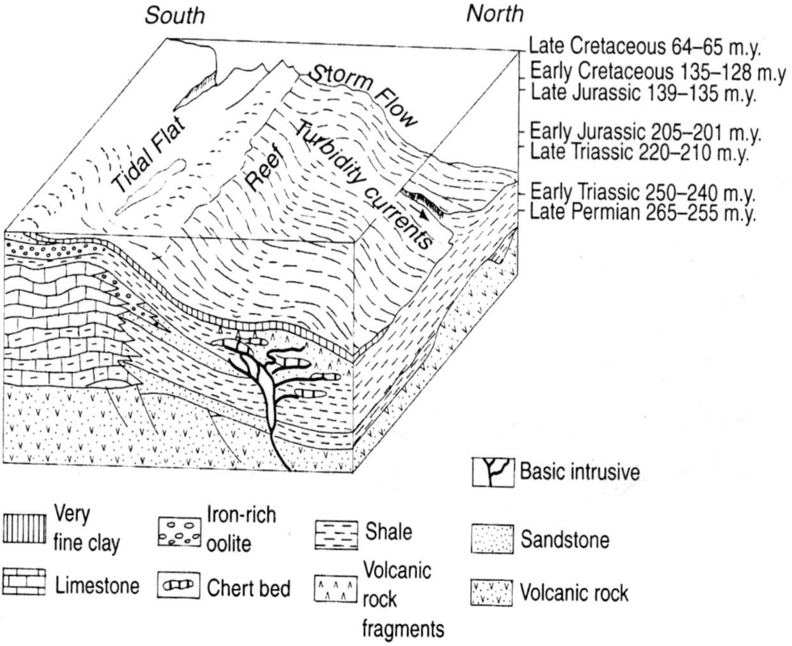


**Figure 5.9** The northern margin of the Indian continent started sinking rapidly. Submarine landslides and turbidity currents were triggered on steepened slopes and lava welled out through fissures that had opened up in the sinking sea floor.

flowing rivers carried not only the volcanic detritus, but also the remains of these plants and deposited them in their lower reaches and deltas now recognized in the Kagbeni, Kangpa and Lingshi areas beyond what is at present the Annapurna–Sagarmatha–Kanchanjangha Range (Figure 5.10).

Nearly synchronous with the subsidence of the northern periphery of the Indian continental margin, the southern edge of the Tibetan landmass of mainland Asia also sank. It sank very rapidly; and by the end of the Cretaceous, about 65 million years ago, it had become a more than 2000 m-deep oceanic trench along the margin of the Asian continent (Figure 5.11). Turbidity currents brought sediments from the unstable continental slopes and deposited them in the trench. In the oceanic trench were deposited material brought down from the landmass of Asia. This included the fragments of a variety of rock types and the remains of animals and plants endemic to the Tibetan landmass. These animals and plants were entirely different from those of the Himalayan world of the Indian continent. Fossils of foraminifers, radiolarians, rudists, bivalves, etc., occurring in the sediments of the seafloor and in the oceanic trench indicate the time period of 83 to 65 million years ago.

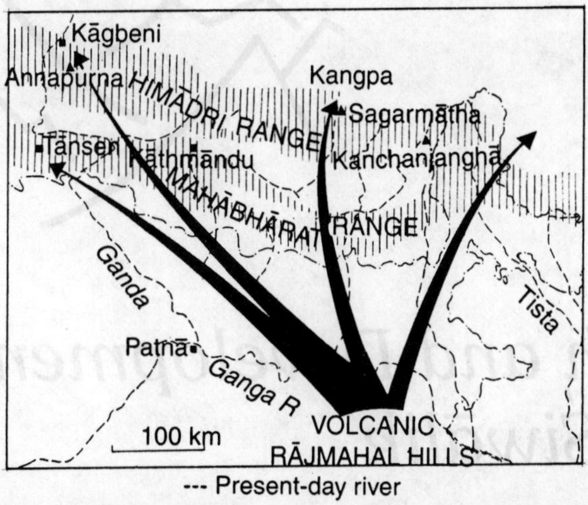


**Figure 5.10** Rivers draining the volcanic terrain of the Rajmahal–Sylhat tract of eastern Gondwanic India carried remains of plants along with volcanic detritus, and deposited them far beyond what is presently the Annapurna–Sagarmatha–Kanchanjangha Range.

Simultaneous with the sinking of the seafloor adjacent to the continental margins, the seafloor developed fissures. Through them poured out lava on a grand scale all along the tract (Figure 5.11). A chain of volcanic seamounts and islands formed in front of the sinking continental margin. The island arc stretched from Kohistan in the northwest, through Kargil in Ladakh and Mansarovar north of Kumaun, to Shigatse in southeastern Tibet. By now the northward-moving Indian land mass had come close to Asia (Figure 5.11). In between the two lay the arcuate chain of the volcanic trenches filled with material derived from the Tibetan landmass. The gap between the converging continents was narrowing, even as the sediment and rock successions of the different domains were getting closer and closer.

Then came the tectonic revolution that culminated in the birth of the Himālaya.

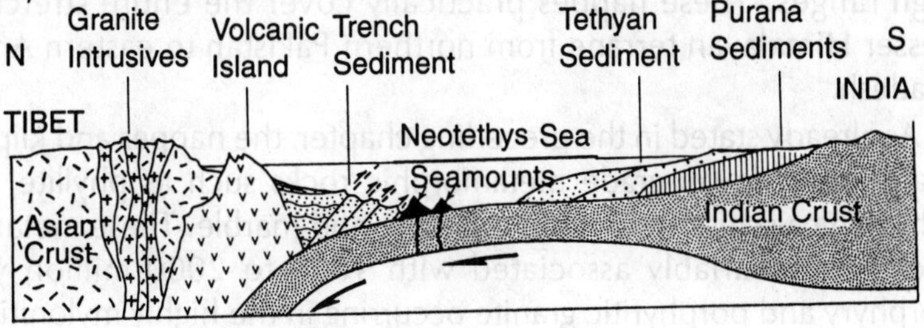


**Figure 5.11** The southern edge of the Tibetan landmass of mainland Asia sank very rapidly to form a > 2000 m deep trench of sorts. The subsiding ocean floor was simultaneously ruptured, and lava came out through fissures to form a chain – an arc – of volcanic islands.

# 6

# *Birth and Development of the Śiwālik*

Shortly after the emergence of the Himādri as a mountain, there was a revival of tectonic movements. The more than 7000 m-thick pile of sedimentary rocks along with overthrust sheets, consisting of metamorphic and granitic rocks in the Lesser Himalayan terrane, was crumpled and thrown into a series of folds. In the north, closer to the foot of the Himādri, the compression and resultant deformation were so strong that the tightened folds were overturned, squeezed out, one toppling over another, and subsequently split by faults along their axial planes (Figure 6.1). Earlier pushed southwards and placed upon the sedimentary rocks, the uprooted rock masses as nappes and klippen were refolded concordantly with the underlying sedimentary rocks. The nappes and klippen (Figure 4.3), occurring securely in the cores of the synclinal folds, escaped subsequent erosion and formed high ranges. These nappes practically cover the entire stretch of the Lesser Himalayan terrane from northern Pakistan to eastern Arunachal Pradesh.

As already stated in the preceding chapter, the nappes and klippen are made up of lower-grade metamorphic rocks such as phyllite, chlorite schist, mica schist, micaceous quartzite and marble. These metamorphic rocks are invariably associated with 1800 to 2000 million year-old porphyry and porphyritic granite occurring in the highly mylonitized and sheared states at the base of the successions, and with the 500 to 540 million year-old rather light coloured granites occurring as sills and batholiths in the upper part of the formation.

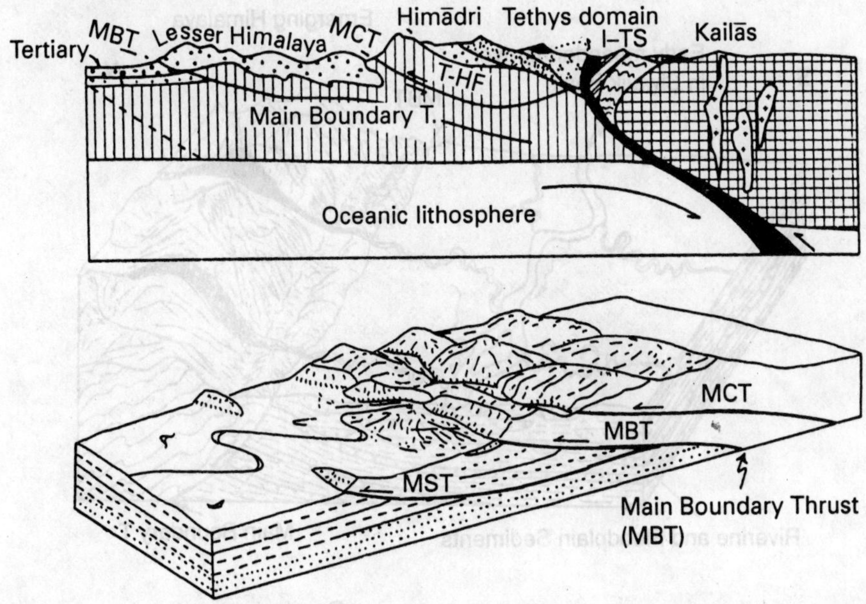


**Figure 6.1** The revival of strong movements on the Main Central Thrust brought the assemblage of the metamorphic and granitic rocks of the Himadri terrane onto and over the sedimentary succession of the Lesser Himālayā. So strong was the deformation below in the MCT zone that the tightened folds were overturned and split repeatedly along their axial planes. In the south the Main Boundary Thrust had developed. (The HFF was yet to form). The block diagram is modified after T Tokuoka and others, 1994.

## Revival of Movement in the Southern Lesser Himālaya

The southward-thrust or displaced rock piles encountered strong resistance in their forward translation. It was maximum all along the 2400 km-southern front, where the crust had earlier ruptured to form the incipient Main Boundary Thrust (MBT). Now, there was a revival of tectonic movement on the multiplicity of faults of the MBT zone. Some of these faults were parallel to the main plane of rupture and others branched off from it. Consequently, slabs and slices of rockmasses were thrust up to form a high ridge of the Outer Lesser Himālaya. This high range overlooked the Sirmaur Foreland Basin to the south. Some slabs or sheets advanced southwards onto this basin, overriding and trampling the sedimentary successions of the Subathu and Murree formations, belonging to the Palaeogene period (Figure 6.2).

## Evolution of Śiwālik Basin

Simultaneously, the ground immediately to its south sagged, giving rise to yet another foreland basin. The floor of this basin – it may be recapitulated

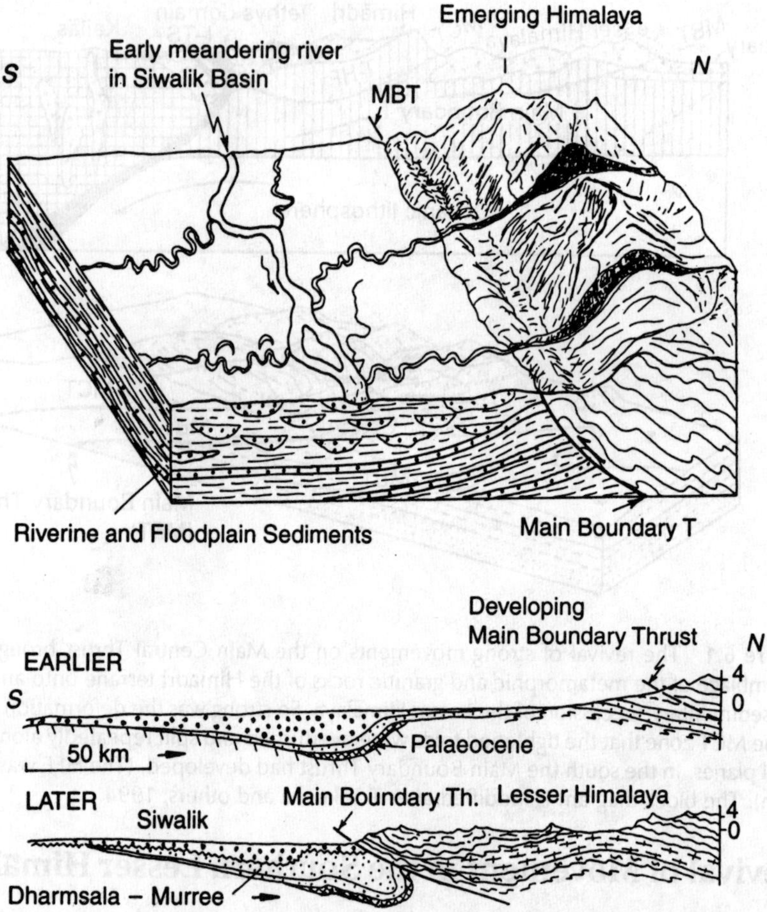


**Figure 6.2** Reactivation of the Main Boundary Thrust, entailing the riding of the Lesser Himalayan rocks on the Sirmaur-Basin sediments, was accompanied by the sagging of the crust immediately to the south of the rising mountain front. The outcome was the development of the Śiwālik Basin—yet another foreland basin.

– was made up partly of the Subathu–Murree succession of the Sirmaur Basin and partly of the Precambrian rocks of Peninsular India. In other words, the new basin spread quite beyond the limits of the Sirmaur Foreland Basin and encompassed a large area stretching from Sindh in southern Pakistan, through the Jammu region in northwestern India, the Śiwālik tract in southern Garhwal, the Churia Hills in Nepal, and the Dihing domain in Arunachal Pradesh to the Tipam–Dupitila region in the Assam–Tripura sector (Figure 6.3). This foreland basin is known as the Śiwālik Basin.

The Śiwālik Basin came into existence 18.3 million years ago. But the main span of sedimentation was 16 to 5 million years ago. The timings

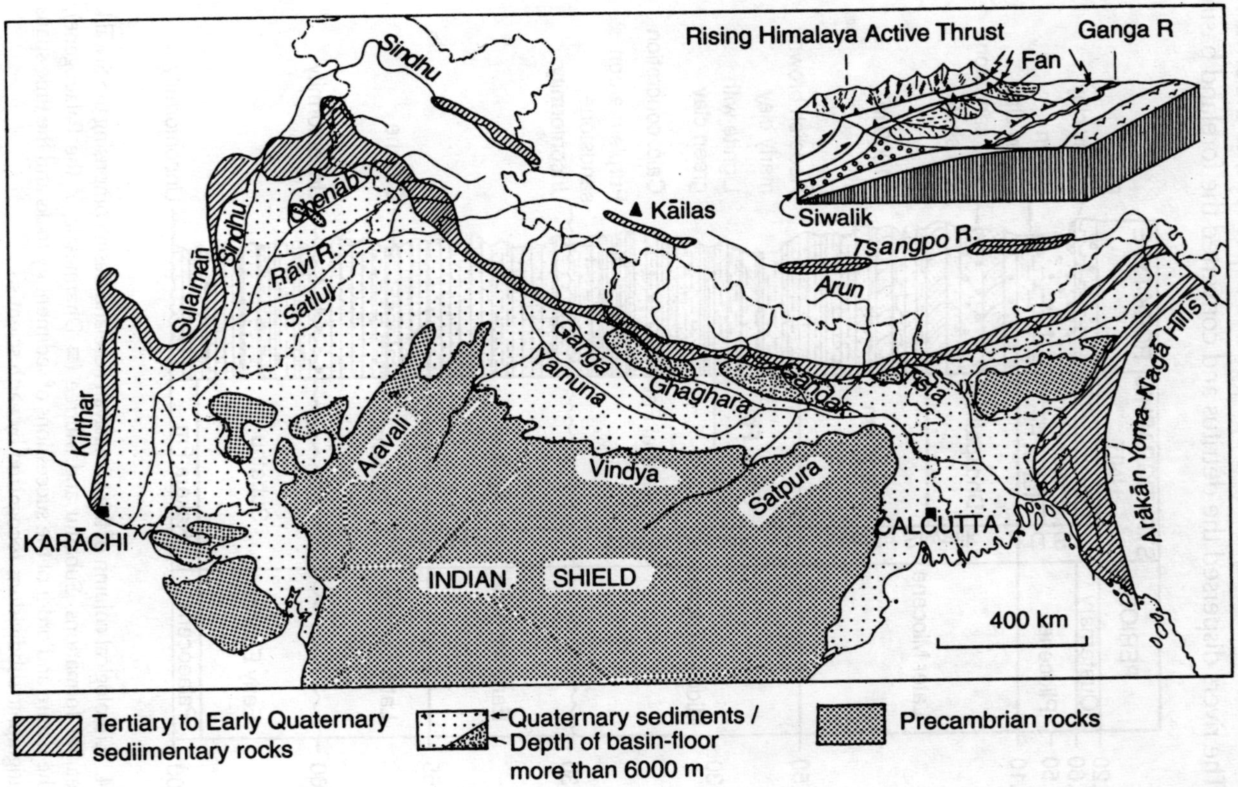


**Figure 6.3** Present-day considerably truncated Śiwālik terrane. The original basin stretched from Sindh in southern Pakistan to Tripura in the east and was much wider than depicted. It is shown delimited by the Himalayan Frontal Fault against the younger Indo-Gangetic Plains that developed much later.

are inferred from the palaeomagnetic–polarity study coupled with fission-track dating of certain heavy minerals in the sedimentary rocks of the Śiwālik. The basin remained a repository of detritus until about 0.22 million years ago, when the process of sediment accumulation ceased. The total thickness of the sediments, preponderantly derived from the now fast-rising Himālaya, is of the order of 7000 m. It was a colossal volume of sediments that the Himalayan rivers dumped and deposited into the Śiwālik Basin. The rivers dispersed the detritus and converted the foreland basin

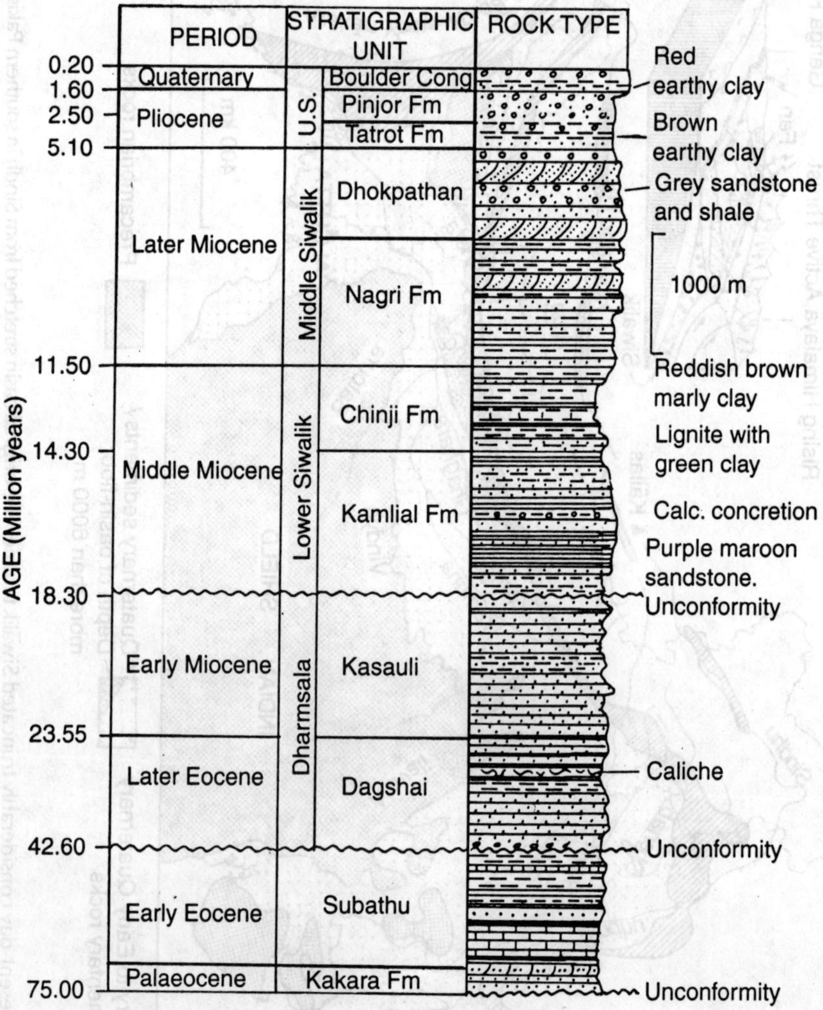


**Figure 6.4** Lithological column of the Himalayan Foreland Basin comprising the Śiwālik and the earlier formations, Subathu and Murree (≡ Dharmasala of the Palaeogene), showing the nature and order of the successions of sedimentary rocks and the time spans of the stratigraphic subdivisions recognized by earth scientists.

into a vast floodplain. They migrated laterally, forming coalescing fans of debris and building multi-storied sand complexes. These rivers were very wide and frequently encroached upon each other's floodplains, as is evidenced by the occurrence of overbank clay and silt.

The Śiwālik succession (Figure 6.4) comprises an alternation of muddy sandstone and maroon mudstone or shale, with occasional beds of pebbly conglomerate (in the Lower Śiwālik), giving way upwards to dominant coarse-grained sandstone that is characterized by micas and calcareous material and interbedded with subordinate grey and brown shale (in the Middle Śiwālik). Towards the upper part, the sandstone becomes more coarse-grained, quite pebbly or even conglomeratic locally (in the Upper Śiwālik). The meandering rivers in the Lower Śiwālik time were transformed into braided rivers in the later Śiwālik epoch—obviously due to steepening of the slope of the alluvial land. This fact implies that the mountain province and its foreland had once again started rising up in the period 5.3 to 0.22 million years ago of the Upper Śiwālik. This is further borne out by the speeding up of the rate of sediment accumulation. This subject is discussed in greater detail in the next chapter.

The rate of sediment accumulation varied from sector to sector of the floodplains and over time. It was 10 cm/1000 yr in the Lower Śiwālik, 30 cm/1000 yr in the Middle Śiwālik of the Potwar region, and 21–70 cm/ 1000 yr in the Upper Śiwālik of the Jammu sector. In Himachal Pradesh the rate was 30 to 40 cm/1000 yr during the Middle Śiwālik epoch and 40 to 50 cm 1000 yr in the Upper Śiwālik time. In Nepal the average rate was 33 cm/1000 yr, the peak of 48–50 cm/1000 yr being in the interval 9 to 10 million years ago.

Heavy minerals, occurring characteristically but in very small quantities in the Śiwālik sedimentary rocks, came largely from the rising Himālaya. The appearance of garnet and staurolite in the Lower Śiwālik at the base indicates that the lower-grade metamorphic rocks had been exhumed – uplifted and exposed to erosion – in the Lesser Himālaya during the period 18.3 to 11.0 million years ago. And the characteristic presence of kyanite with epidote and staurolite in the Middle Śiwālik implies the uplift and denudation of higher-grade metamorphic rocks that make up the bulk of the Himādri or Great Himālaya. Obviously, in the period 11.0 to 5.3 million years ago, the Himādri had become a high mountain and quite vulnerable to erosion. The high-grade sillimanite in the Upper Śiwālik rocks indicates that by the time 5.3 to 1.6 million years ago even the deepest rocks of the Himādri domain had been uplifted and exposed to denudation. To put it differently, the Himālaya mountain grew progressively higher and higher in the period 18.3 to 1.6 million years ago of the Middle to Upper Miocene period.

## Climatic Conditions and Life in the Early Śiwālik Time

It was a warm humid climate that prevailed during the Lower Śiwālik time 18 to 11 million years ago. This is evident from the nature of red sediments and the testimony of fossils of plants and animals that were buried in the sediments of the Śiwālik floodplains. The floodplains were covered thickly with forests of tropical evergreen trees (Figure 6.5A). With the abundance of food available in these rainforests where water was aplenty, a rich and varied life grew and flourished in the land of rivers, swamps and lakes. In the rainforests of the Lower Śiwālik time pigs were abundant in the marshy tracts along with a variety of elephants, carnivores and artiodactyls. Among the bulky animals that roamed the forests of the Lower Śiwālik time was the tree-top browser, the hornless giant *Baluchitherium* of the rhino family. First discovered in Baluchistan, this was the largest land mammal that ever lived on earth (Figure 6.7). Among the contemporary creatures was the ape *Sivapithecus* (Figure 6.6A) that lived in open–mixed woodlands 14 to 10 million years ago, in the Śiwālik forests of Ramnagar (in Jammu)—Hartayalnagar (in the Himachal) belt. The *Sivapithecus* (also described as *Ramapithecus*) bore facial and dental resemblence to the orangutang that lives at present in southeast Asia. From the fossils of teeth and jawbones recovered from the red mud of the Śiwālik belt in northern Pakistan, it appears that *Sivapithecus* was a primate somewhat larger in size than the male orangutang. These primates lived in the dense forests dominated by *Dipterocarpus* and *Calophyllum* trees (Figure 6.5B). A variety of reptiles including the giant turtle, *Colossochelys atlas,* (Figure 6.6B) lived in the water bodies.

At the time when life flourished in the Śiwālik terrane, the Himālaya was growing taller and taller even as rivers worked furiously at sculpturing the terrane and fashioning the landscape. The topography of the Himālaya that we see today is largely the outcome of the changes that took place in this period. Needless to state, later tectonic movements and attendant changes considerably modified that geomorphological layout.

(A)

(A)

**Figure 6.5** (A) Forests in the Lower Siwalik time provided food and water aplenty to a wide variety of animals including the quadrupeds, some of which were very large. *(Photo: A C Nanda)* (B) Leaf impression of *Dipterocarpus* that dominated the Lower Siwalik forests *(Photo: A K Sinha)*

(A)

(B)

**Figure 6.6** Life in the Early Siwalik time (A) Primate *Sivapithecus*, also known as *Ramapithecus*, which bore some resemblance to the modern *orangutang* in southeast Asia (Sketch by Sudip K Pal) (B) Giant turtle *Colossochelys atlas*. *(Photo: A C Nanda)*

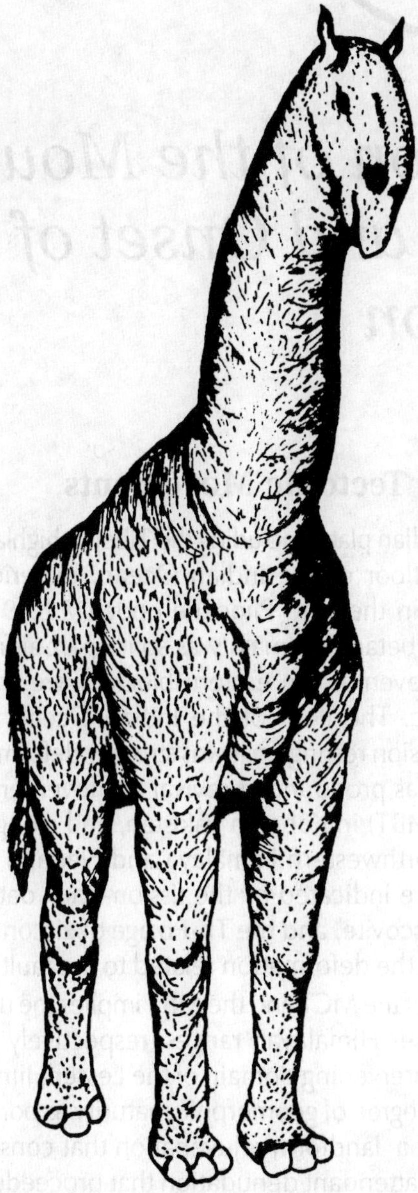

**Figure 6.7** The hornless giant *Baluchitherium* was a tree-top browser that lived in the Śiwālik terrane of Baluchistan, 10–14 million years ago. *(Sketch: S K Paul)*

# 7

# *Evolution of the Mountain Barrier and Onset of the Monsoon*

## Late Miocene Tectonic Movements

Large parts of the Indian plate, including the Tibetan highland, the Himalayan province and the floor of the Indian Ocean, experienced a powerful tectonic upheaval in the Late Pliocene epoch 7 to 9 million years ago (Figure 7.1). The Tibetan landmass was split apart along N–S faults $8 \pm 1$ million years ago, even as the eastern blocks were pushed further east towards the Pacific. The floor of the Indian Ocean together with the sedimentary succession resting on it was strongly deformed 7.5 to 8 million years ago. There was pronounced revival of movements along the Main Boundary Thrust (MBT) in Kohat in Pakistan, and along the Main Central Thrust (MCT) in northwestern Himālaya and Nepal 6 to 8 million years ago. These dates are indicated by the fission-track dating of apatite, the cooling time of muscovite, and the Th-Pb age of zircon occurring in rocks that were caught in the deformation related to the fault reactivation.

The reactivation of the MCT and the MBT implies the uplift of the Himādri and the Outer Lesser Himalayan ranges, respectively. Between the two uplifted belts, the intervening domain of the Lesser Himālaya – which had by then attained a degree of geomorphic maturity as borne out by a mantle of soil – underwent a landform rejuvenation that considerably reshaped the landscape. The attendant denudation that proceeded inexorably with heightened vigour, generated great volumes of detritus, and rivers and streams carried them to the Śiwālik Basin in the south.

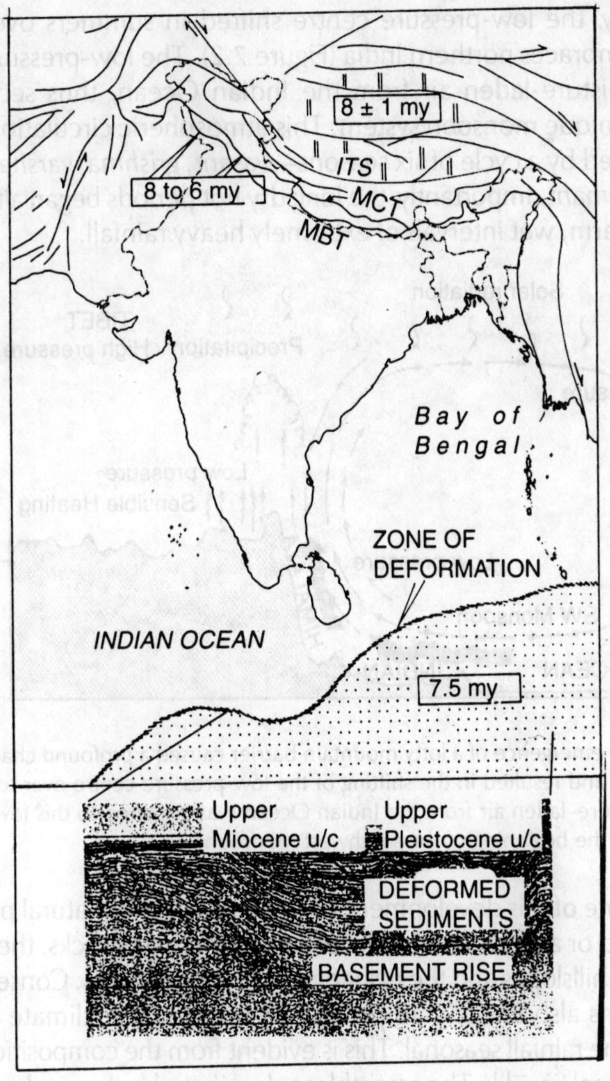


**Figure 7.1** The Tibetan landmass experienced extension, the MCT and MBT were reactivated in the Himalayan province, and the Central Indian Ocean was affected by strong deformation (inset) in the short interval 7 to 9 million years ago. Arrows indicate block movement and vertical lines depict faults.

## The Barrier

Strong and possibly repeated movements along the MCT culminated in the uplift of the Himādri to such a height that it became an insuperable barrier. This brought about a profound change in the atmospheric circulation. The barrier disrupted the then west-to-east flow of the winds and caused northward displacement of the high-pressure belt.

Consequently, the low-pressure centre shifted in summers over to the belt which embraces northern India (Figure 7.2). The low-pressure centre attracted moisture-laden air from the Indian Ocean, thus setting into motion the unique monsoon system. This atmospheric circulation system is characterized by a cycle of six seasons—*vasant*, *grishma*, *varsha*, *sharad*, *shishir* and *hemant*. Importantly, the long dry hot periods began alternating with short, warm, wet intervals of extremely heavy rainfall.

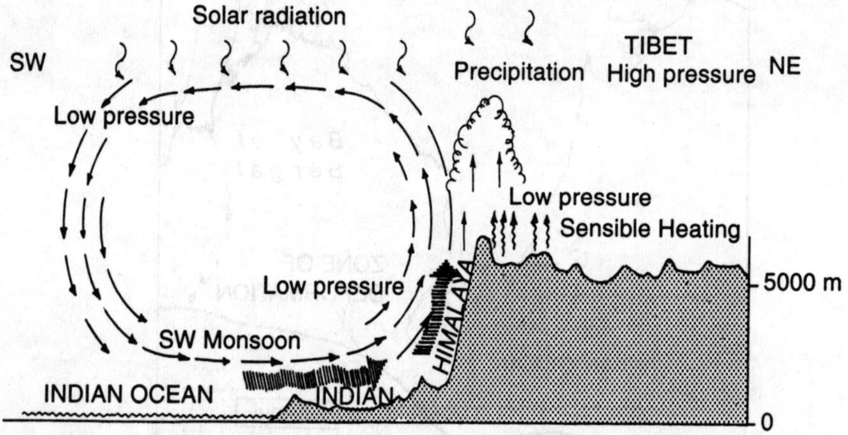


**Figure 7.2** The emergence of a lofty mountain barrier caused a profound change in the wind circulation and resulted in the shifting of the low-pressure centre over to northern India. The moisture–laden air from the Indian Ocean was attracted to this low pressure centre. This was the beginning of the southwest monsoon.

The outcome of this development was that a variety of natural processes were activated or accelerated, such as the weathering of rocks, the erosion on steepened hillslopes and the reshaping of the topography. Consequently, the ecosystems also underwent profound changes. The climate became warmer and the rainfall seasonal. This is evident from the composition of the flora in the Nepal Śiwālik. The assemblage dominated by tropical evergreen large trees such as *Dipterocarpus*, *Shorea*, *Hopea*, *Calophyllum*, etc., gave way to the assemblage in which there was great abundance of deciduous vegetation including *Bauhinia*, *Terminalia* and tall grasses. Strong variations in the carbon-isotope values of carbonates in the Śiwālik soil corroborate the above-mentioned vegetation change 6 to 7.4 million years ago in northern Pakistan, and 7 to 8 million years ago in southeastern Nepal.

## Churning in Indian Ocean

What happened in the Indian Ocean 7.4 to 8.5 million years ago testifies to the profound climatic changes that had taken place in this part of the globe. There was, rather suddenly, a prolific growth of tiny floating and swimming animals in the high sea—the foraminifer *Globigerina bulloides* and the radiolarians *Collosphaera* aff. *huxleyi* and *Actinomma* became very prolific in the warm water in that brief duration (Figure 7.3A). This development signifies the abundant availability of nutrients in the surface water. In the high seas, only the upwelling currents could have swept up the nutrients from the ocean bottom. And the upwelling currents must have been set in motion by the atmospheric currents above the ocean water. Evidently, 7.4 to 8.5 million years ago, a new system or pattern of atmospheric currents flowing northwards from the Indian Ocean was established. This was the onset of the southwest monsoon.

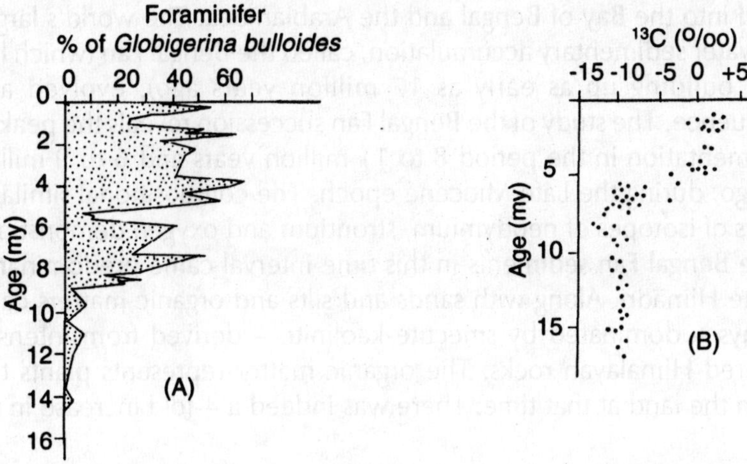


**Figure 7.3** (A) Prolific growth of tiny swimming and floating animals – such as the foraminifer *Globigerina bulloides* – in the high seas 7.4 to 8.5 million years ago, implies abundant availability of nutrients in the surface water. Only the upwelling currents sweeping the ocean-floor could have brought up the nutrients from the ocean bottom. (B) Organic matter derived from the soils of the Himalayan terrane increased suddenly at 9 million years, substantially at 6.5 million years, and very greatly at 0.85 million years ago. The significance of the latter two dates is discussed in later chapters.

## Acceleration of Erosion

With the heavy rains of the southwest monsoon relentlessly beating upon the uplifted Himādri, it was but natural that the pace of erosion increased considerably. This is evident from the 2½-fold increase in the rate of sedimentation in the Śiwālik Basin. This happened 11 million years ago in the Potwar sector in Pakistan, and 9 and 10 million years ago, in southeastern Nepal. There was a five-fold increase in sediment influx into the Bay of Bengal in the interval 7 to 11 million years ago (Figure 7.4). Together with the sediments came some very diagnostic heavy minerals, which reveal their parentage in the Himālaya mountain. The mineral kyanite and sand-sized grains of gneiss derived from the deeper-level rocks of the Himādri, appeared 9.2 million years ago in the Śiwālik sediments of Nepal. The occurrence of diagnostic kyanite and sillimanite together with calcic amphibole, in the middle and upper parts of the sedimentary succession of the Śiwālik, imply that way back in the period 5 to 11 million years ago the deep-seated rocks of the Himādri terrane were exhumed—uplifted and exposed to strong erosion.

So voluminous was the influx of detritus derived from the rising mountain that the excess sediment spilled over the Śiwālik Basin and entered into the Bay of Bengal and the Arabian Sea. The world's largest underwater sedimentary accumulation, called the Bengal Fan (which had started building up as early as 17 million years ago), evolved as a consequence. The study of the Bengal Fan succession reveals the peaking of sedimentation in the period 8 to 11 million years and 6 to 9 million years ago, during the Late Miocene epoch. The compositional similarity in terms of isotopes of neodymium, strontium and oxygen demonstrates that the Bengal Fan sediments in this time interval came predominantly from the Himādri. Along with sands and silts and organic matters came fine clays – dominated by smectite-kaolinite – derived from intensely weathered Himalayan rocks. The organic matter represents plants that grew on the land at that time. There was indeed a 4-fold increase in the

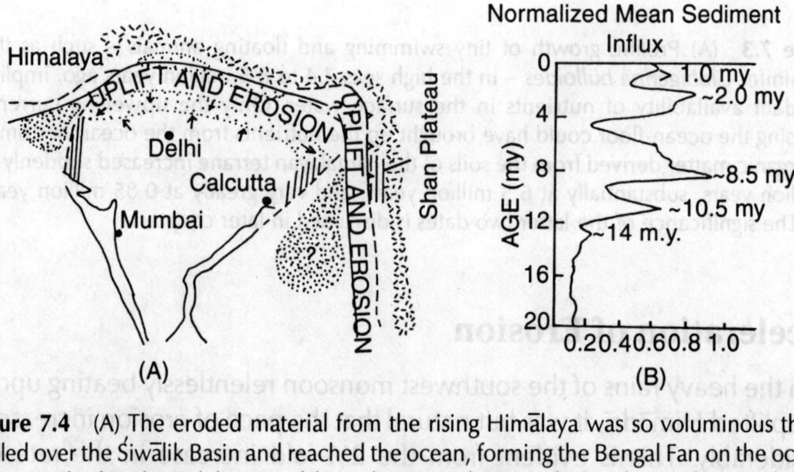


**Figure 7.4** (A) The eroded material from the rising Himālaya was so voluminous that it spilled over the Śiwālik Basin and reached the ocean, forming the Bengal Fan on the ocean-floor. (B) The fan-shaped deposit of the sediment in the Bay of Bengal records several peaks of heavy sedimentation including in the interval 11 to 8 million years ago. High sediment influx means stronger erosion in the provenance—in the original site of the concerned rocks. This must be due to heavy monsoon rainfalls.

influx of land-derived organic matter into the northern Indian Ocean about 9 million years ago and later at 6.5 million years ago (Figure 7.3B). This fact points to the existence of terrestrial vegetation nurtured by the soil covering the Lesser Himalayan terrane that was then exposed to brisk erosion.

Needless to emphasize, the later part of the Upper Miocene epoch (7.5 to 11 million years ago) witnessed accelerated erosion in the Himalayan province, particularly in the Himādri domain which had by then risen up as a lofty rampart.

## Invasion of Exotic Animals

In the wake of the onset of the monsoon, grasslands began encroaching on tropical rainforests in the Śiwālik floodplain. The long, dry hot periods followed by short, warm wet spells were conducive to the development of

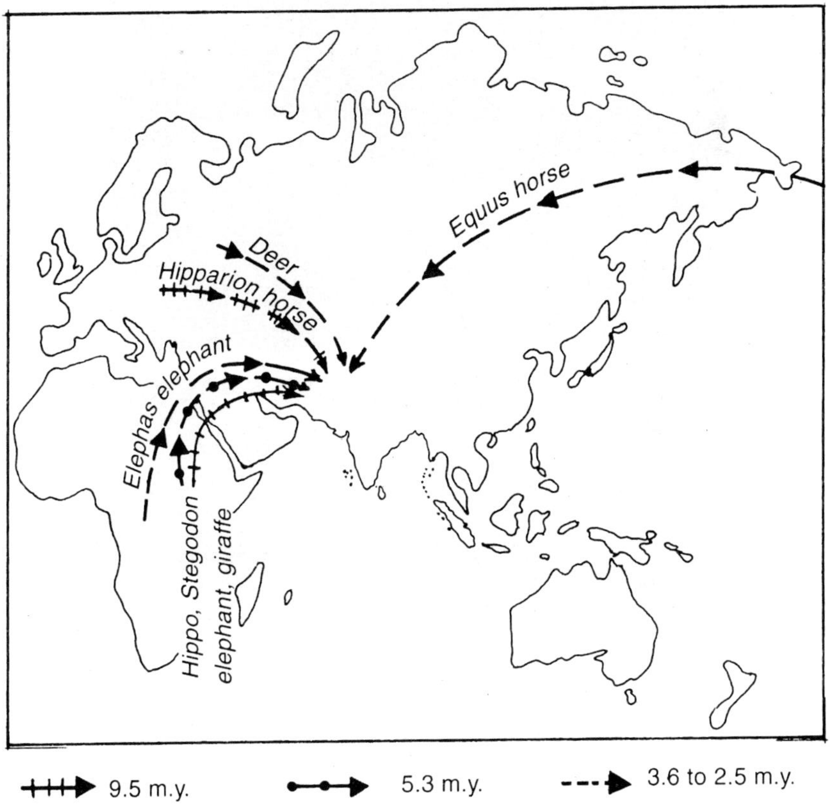


**Figure 7.5** Attracted by the grasslands in the Śiwālik terrane, quadrupeds immigrated to the land of plenty from neighbouring lands. They came in several waves through routes across the then rising Himalayan province. The Himādri had already become a lofty barrier.

(A)

(B)

(C)

**Figure 7.6** Immigrants from distant lands. (A) *Hipparion*, the three-toed horse, came from Europe 9.5 million years ago (B) *Hexaprotodon*, the hippopotamus came from Africa 5.3 million years ago and (C) *Elephas planifrons,* the elephant, followed suit 3.6 million years ago. *(Photos: A C Nanda)*

savannah-type forests and grasslands. Among the trees, the pines (*Pinus*) had secured a foothold in the forests. The expansion of grassland with its rich resources of forage and food attracted grazing animals from the neighbouring lands. They came to the land of plenty – the Śiwālik – herds after herds, through different routes across the newly-emergent Himālaya (Figure 7.5). The immigration of exotic quadrupeds brought about a major faunal turnover in the Śiwālik world. A notable change occurred 7.5 to 9.5 million years ago in the Potwar region in northern Pakistan. The introduction of exotic fauna in large numbers led to the inevitable marginalization – and in some cases, extermination – of the indigenous animals such as rhinos, buffaloes and cows.

Of all the animals that came to the Śiwālik domain, the three-toed horse *(Hipparion)* was possibly the pioneer immigrant (Figure 7.6). It came from Europe via Central Asia 9.5 million years ago. At the same time, the boar (*Suid*) came from Africa. *Solenoportax* came about 7.4 million years ago from Africa, the proboscid elephant *(Stegodon)* and the hippopotamus *(Hexaprotodon)* arrived around 5.3 million years along with the langur (*Presbytis)* the carnivore, *Dinofelis,* and the giraffe *(Vishnutherium)* from central Africa. Quite some time later, about 3.6 million years ago, the elephant (*Elephas planifrons)* came from Africa; and still later, about 2.5 million years ago, came the one-toed horse *(Equus)* all the way from Alaska in North America. Almost at the same time, the deer (*Cervus*) immigrated from Central Asia. The animals that made the Śiwālik their home proliferated and diversified very rapidly in the congenial environment of the floodplains. Understandably, the Śiwālik faunal assemblage of that time was three times richer than what it is today.

The Late Miocene immigration in large numbers of four-footed animals across the very youthful Himalayan province, implies that some segments of the mountain belt were simply not high and rugged enough to impede or prevent movements of those bulky beasts 7.4 to 9.5 million years ago. Admittedly, a few corridors could have been enough for free immigration. However, considering the large size of their bodies, and their wide distribution in the Śiwālik, in the Lesser Himālaya and in contemporary Tibet, more than corridors would have to be invoked to explain the heavy influx of these animals. Perhaps, the whole of the western flank was open when these animals entered the Indian subcontinent (Figure 7.5).

## Life in Upper Śiwālik Time

In the Upper Śiwālik time, 5.1 to 1.6 million years ago, the environmental conditions had changed considerably—the tropical forests were largely replaced by savannah-type, grassy plains dotted sparsely with trees. Grazing and browsing animals became preponderant in the forests dominated by *Terminalia* ('sain'), *Mangifera* ('mango'), *Bauhinia* ('kachnaar'), *Albizzia* ('sirees'), *Acacia* ('khair'), etc. In these forests lived the *Macacus* monkeys and the *Simia* and *Semnopithecus* apes; the carnivores tigers, hyaenas, panthers and cats; the elephants *Mastodon sivalensis*, *Stegodon ganesa*, *Elephas planifrons* and *Elephas hysudricus;* the giraffes *Indratherium* and *Sivatherium;* the ungulates rhinos, horses, hippopotami, boars and camels; and the artiodactyls deer, buffaloes, cows and bisons.

The presence of stone artefacts in the lower sediments of the Upper Śiwālik succession in the Suketi valley in southeastern Himachal Pradesh provides clues to the presence of human-like primates in the Śiwālik floodplain. However, no body remains have so far been found anywhere in the Śiwālik terrane.

## Formation of Intermontane Lakes

In the Later Pliocene epoch, roughly 4 to 2 million years ago, many parts of the Himalayan province and Tibet experienced crustal stretching as they were split by faults. Synchronously, the floor of the Indian Ocean was subjected to strong deformation. The rupturing of the land in the

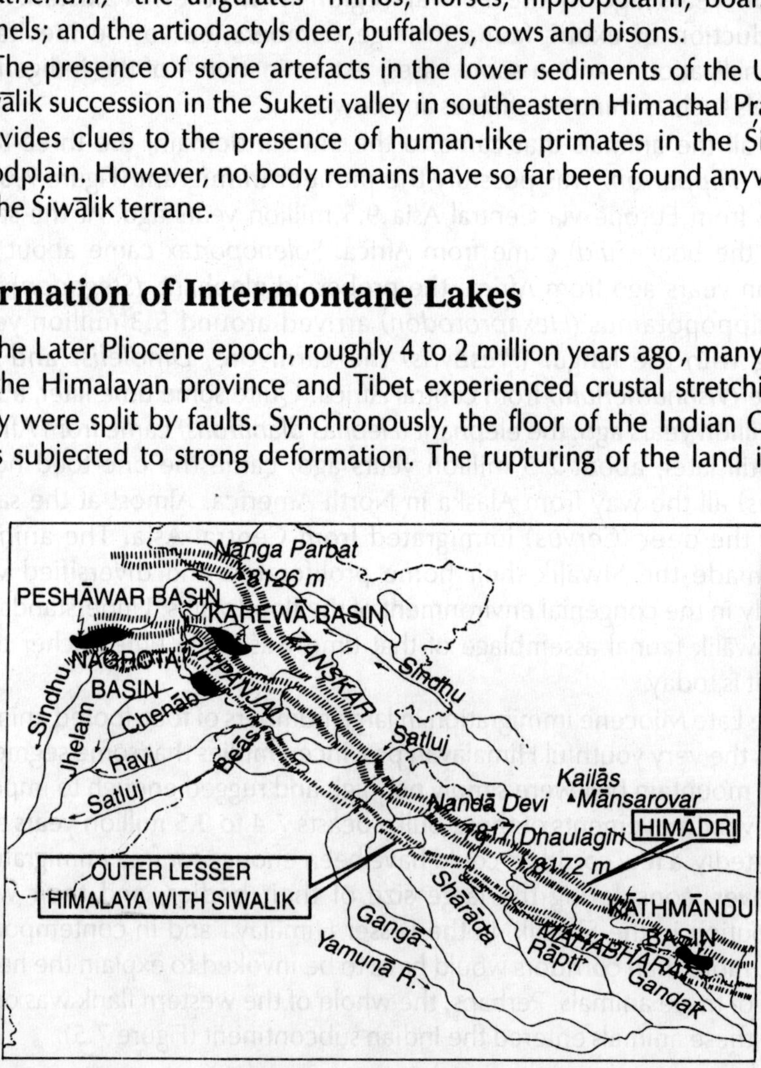


**Figure 7.7** The uplift of the downstream block caused the ponding of rivers and formation of lakes upstream in the river valleys—the Karewa in Kashmir, the Peshawar in Pakistan, the Nagrota in Jammu, and the Kathmandu in Nepal. These lakes are at present separated by formidably high mountain ranges (lakes are shown in black).

northern part of the Himālaya and adjoining Tibet caused east–west extension, giving rise to N–S oriented faults and the formation of grabens, such as the Gulu north of Bhutan and the Thakkhola in north–central Nepal along the Kali Gandaki Valley.

Fission-track dating of apatite, zircon and hornblende indicate that there were concommitant movements along the boundary faults like the T-HF, MCT, MBT, and others related to them. Movements along faults involving uplift of the downstream blocks, caused the blockage of streams and rivers and the resultant formation of lakes in the river valleys (Figure 7.7). In Kashmir the faulting up of the PirPanjal Range caused the ponding of the Jhelam River, and the formation of the Karewa Lake between the Zanskar and PirPanjal Ranges. The Karewa Lake is represented by nearly 1300 m-thick, 4 to 0.4 million year-old succession of silt and clay with occasional layers of gravel emplaced by landslides and attendant debris avalanches (The top of the lake sediment is now covered by a 25 m-thick mantle of wind-borne silt called loess). Yet another lake developed in northern Pakistan when the Sindhu and its tributary rivers Kabul and Swat were blocked 2.8 million years ago, following the uplift of the Attock Range downstream. The 300 m lacustrine succession forms the plain of the Peshawar Basin.

In central Nepal, where the Bagmati River was blocked 2.8 million years ago due to the uplift of the Mahabharat Range along faults of the MBT zone, a big lake was formed in what is today the Kathmandu Basin. The lake existed until about 1 million years ago. A similar ponding of the Kali Gandaki River in its uppermost reaches in the Thakkhola Graben resulted in the lake formation, the deposits of which occur within the >850 m succession of the graben. The age of the fluvial sediments of this graben ranges from earlier than 3 to 2.5 million years ago.

## Testimony of Lake Fossils

The surface of the 300 m-thick lake succession of the Kathmandu Basin is today at an altitude of 1300 to 1500 m above sea level. The top of the 1300 m-thick Karewa succession is 1700 to 1800 m above sea level. Within the sedimentary succession of these lakes occur remains of, among others, such animals as the elephants *(Elephas planifrons* and *Stegodon ganesa)* and the hippopotamus *(Hexaprotodon)*. South of the presently 4100 to 3600 m-high PirPanjal Range lies the Upper Śiwālik Nagrota Basin in Jammu (Figure 7.8). Fossils remarkably similar to those of the Karewa also occur in the Nagrota succession. The crocodile *(Crocodylus)* lived in both the Nagrota and the Kathmandu basins.

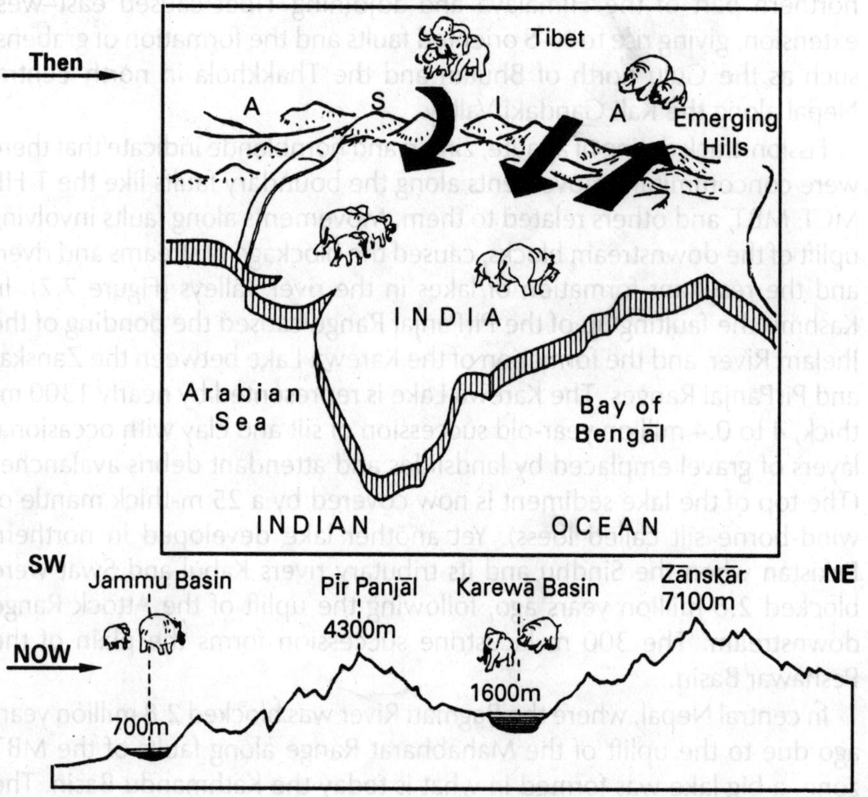


**Figure 7.8** Hippos and elephants could move freely across the PirPanjal terrain from the Karewa Basin to the Nagrota Basin, when they lived 4 to 2 million years ago. Since then, the PirPanjal has risen high to form an insuperable barrier between the two basins.

The striking resemblance – implying a strong biological affinity – of the animals of these widely-separated basins, across what are today very high mountains (Figure 7.7), suggests that even large-bodied and heavy-footed animals could at that time move freely from basin to basin. Possibly, when they lived 4 to 2 million years ago, there were no high mountains across their paths of movement—no rugged tracts to impede or prevent their migration (Figure 7.8).

# 8

# *Collapse of the Mountain Front and Formation of the Indo-Gangetic Basin*

## Tyranny of Tectonic Turmoil

The tectonic upheaval of exceptional severity, that occurred approximately 1.6 million years ago, convulsed the southern ranges of the Lesser Himā laya and the Śiwālik domain. It was the zone of the Main Boundary Thrust (MBT), which had become the main theatre of structural disturbance and widespread natural hazards. As the Outer Lesser Himālaya heaved up along the MBT, its southern front virtually collapsed all through its stretch. Stupendous landslides (Figure 8.1) ravaged the steepened hillslopes, and debris flows carried down voluminous detritus to the Śiwālik Basin. The outcome was the emplacement of roughly 1800 to 2800 m-thick deposit of gravel and mud—described as the Upper Śiwālik Boulder Conglomerate (Figure 8.2). The imposing horizon of gravel is traceable from the Potwar Plateau in the northwest to the Dihing Valley in the east. Great masses of sediment comprising of pebbles, cobbles, boulders and mud were brought down by flooded rivers in great fury. The emplacement took place very swiftly, the rate varying from 21–71 cm/1000 yr in the Jammu sector to 40–50 cm/1000 yr in the Himachal area.

Volcanoes had become active and were spewing fumes and ashes in faraway Afghanistan, Tibet and Myanmar. The wind carried the volcanic ash and shed it on the Śiwāalik domain. It occurs intercalated with conglomerates in the form of tuff, now converted into bentonitic clay. The volcanic ash suggests a date of 1.61±0.1 million years (There were other earlier volcanic events also at 9.46±0.59 million years and 3.0 and later at

(A)

(B)

**Figure 8.1** Collapse of the mountain front – like those at present in south-central Kumaun and northcentral Garhwal – in the zone of the Main Boundary Thrust and Main Central Thrust, respectively, was responsible for the generation of stupendous volumes of gravel and mud. *(Lower photo: K S Bisht)*

(A)

(B)

**Figure 8.2** Emplacement of voluminous masses of debris by flooded rivers and debris flows gave rise to the Upper Siwalik Boulder Conglomerate at the time roughly 1.6 to 0.22 million years ago. The Upper Siwalik Boulder Conglomerate associated with mud in the valley of the Ramganga in south-central Uttaranchal. (Photos: T Tokuoka)

1.5 million years ago). The volcanic ash alludes to the timing of the tumultous tectonic event that overtook the Śiwālik terrane towards the terminal stage of its development. Fission-track dates given by apatite and zircon from the rocks in the zones of deformation corroborate the reactivation of the Main Boundary Thrust, Main Central Thrust and Trans-Himādri Fault, in the interval 1.5 to 2.0 million years ago, within the Himalayan province. Clearly, the whole of the Himālaya suffered the tyranny of tectonic movements at the turn of the geological period from Tertiary to Quaternary, 1.6 million years ago.

## Structural Deformation and Displacement

As a consequence of the reactivation of the MBT, the stupendous pile of the Lesser Himalayan rocks together with the dislodged parts of the sedimentary succession of the Sirmaur Basin moved southwards onto the Śiwālik rocks. The displaced rock masses buried and trampled the Śiwālik rocks underneath them. In the proximity of the MBT the severely-tightened folds were split by faults along their axial planes, and in many places were thrust southwards. The intensity of deformation diminished progressively southwards. It is this deformation of the Sirmaur–Śiwālik piles of sedimentary rocks that gave rise to a variety of structures, serving as traps of petroleum oil and gas (Figure 8.3). Not only the Śiwālik terrane but also the India-Asia collision zone experienced this compressive deformation. This is evident from the folding and northward-thrusting of the Kailas Conglomerate in the Ladakh and Kailas–Mansarovar tracts (Figure 2.5).

## Evolution of the Abode of Snow

A study of the fan-shaped accumulation of sediments in the Bay of Bengal demonstrates that suddenly, in the brief interval 0.8 to 0.9 million years ago, the rate of sedimentation increased greatly. Not only was there an increased influx of sediments, but there was also a predominance of illite and chlorite clays, together with an excess occurrence of organic matter of terrestrial origin, at a point of time 0.85 million years ago. This fact implies that the Himalayan terranes covered with soil containing organic matter (humus) admixed with clays, was uplifted and exposed to denudation. Incidentally, this also implies that the terrane was clothed with vegetation 850,000 years ago (Figures 8.4A and B).

At almost the same time the floor of the Central Indian Ocean suffered strong deformations, as borne out by the structures formed in the overlying sediments. Southeastern Tibet was once again torn apart along faults during the same period. Obviously, the whole of the Indian plate was

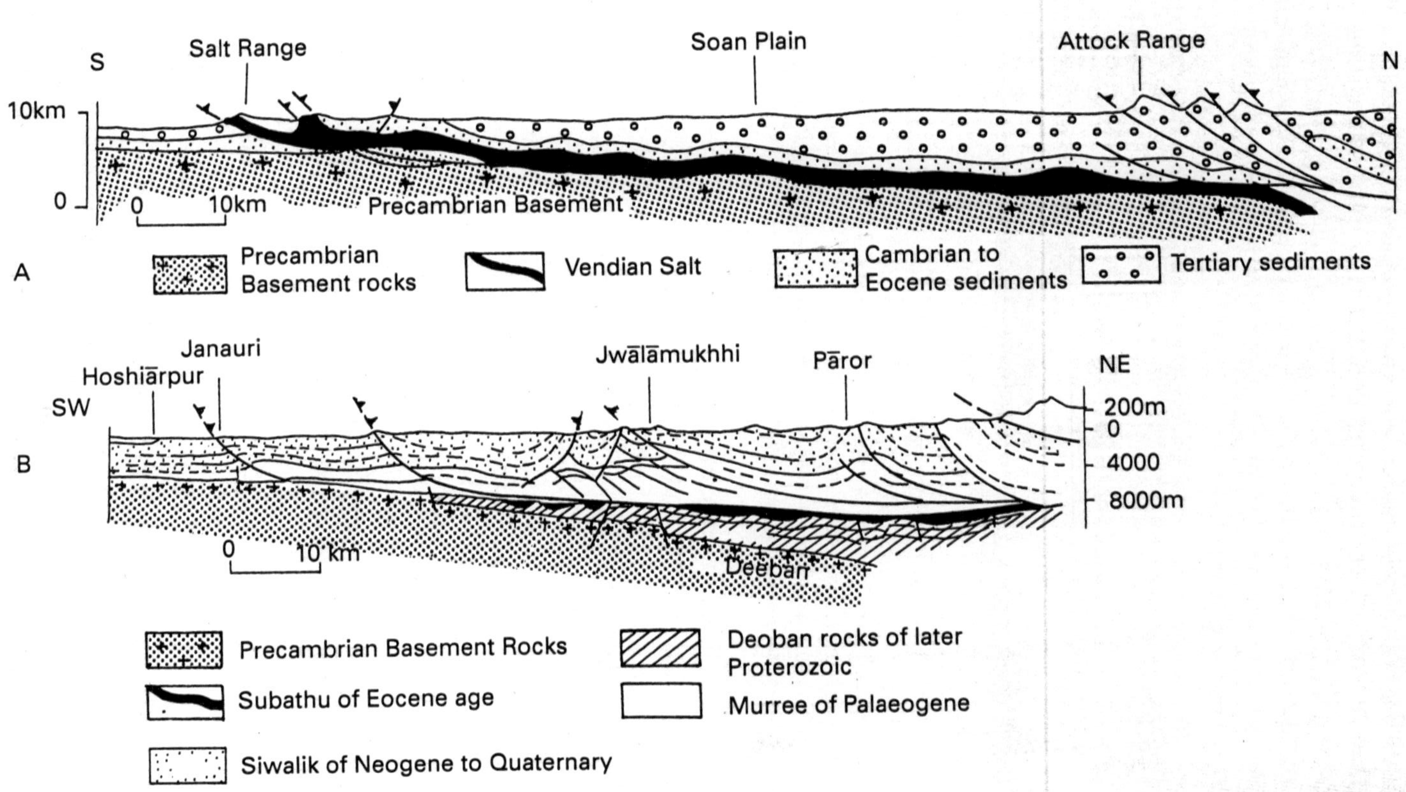


**Figure 8.3** Cross-sections of the Śiwalik terrane demonstrating its structural architecture. The intensity of deformation diminishes southwards, away from the MBT. It may be noted that the Subathu and equivalent rock formations occurring beneath the pile of sediments is the source of petroleum oil and gas that later migrated upwards and were trapped in the structures formed in the younger rocks.

under tremendous strain 0.8 to 0.9 million years ago. Consequently, the Himālaya rose higher than it had earlier been. The uplifted ranges must have diverted the even flow of moist winds, creating large areas of coolness in the higher realm. Cooler conditions conducive to refrigeration thus developed in the high mountains of the Himādri domain, culminating in the onset of glaciation. This was the beginning of the Pleistocene Glaciation in the Himālaya. The "abode of snow" (*Him* = snow + *alaya* = abode or home) had come into existence.

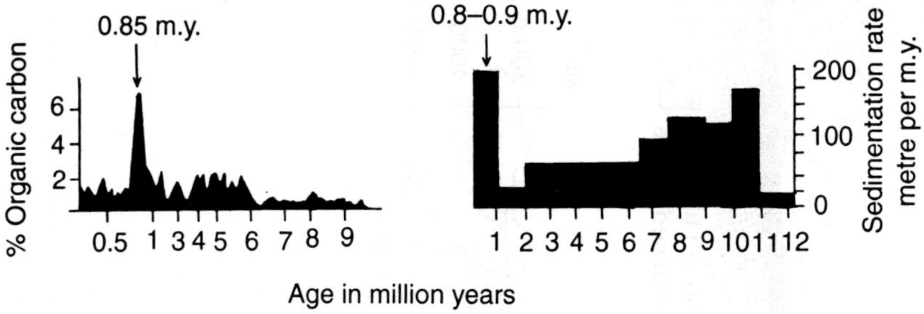


**Figure 8.4** (A) Organic carbon derived from the soils of the Himalayan terranes increased substantially in the Indian Ocean bottom sediments in the period 0.9 to 0.8 million years ago, implying sudden exposure of the Himalayan terranes to denudation. (B) Likewise, the rate of sedimentation greatly and suddenly increased during this period.

In the course of time, the mountain glaciers spread far and wide in Potwar, Kashmir, Ladakh, Tibet and most parts of the Himādri domain (Figure 8.5A). There were, in fact, four successive advances of glaciers, the oldest descending to 1675 m above sea level and the youngest reaching down to 2400 m (Figure 8.5B). The glaciation was not synchronous throughout the region—in some areas the glaciers were most extensive 30 000 to 60 000 years ago and, at other places, 18 000 to 20 000 years before the present. The glaciers left behind great volumes of gravelly detritus in the valleys and also fine-grained sediments, including the varvite (clay-silt alternation) formed in glacial lakes.

The period of glaciation alternated with intervals of warm-wet climatic condition when vegetation grew luxuriously. The plant material occurs in the fossilized soils that are interbedded with gravelly glacial detritus. In a few cases, the vegetal matter was converted to lignite. The palaeosols – soils of the past – recount the history of an altogether different time.

## Partitioning of the Foreland Basin

The tectonic upheaval 850 000 years ago, that was responsible for the uplift of the Himādri ranges, caused the breaking up of the Śiwālik Foreland Basin into two unequal parts (Figure 8.6). The dismemberment

(A)

(B)

**Figure 8.5** The onset of the Pleistocene Glaciation in the Himadri terrane was a consequence of its sudden uplift beyond a critical height. (A) Snow cover today in northeastern Himachal Pradesh and adjoining Ladakh. (B) Glaciers like the Milam (in Kumaun) of the present time filled the valleys in the Himadri domain 0.8 to 9 million years ago.

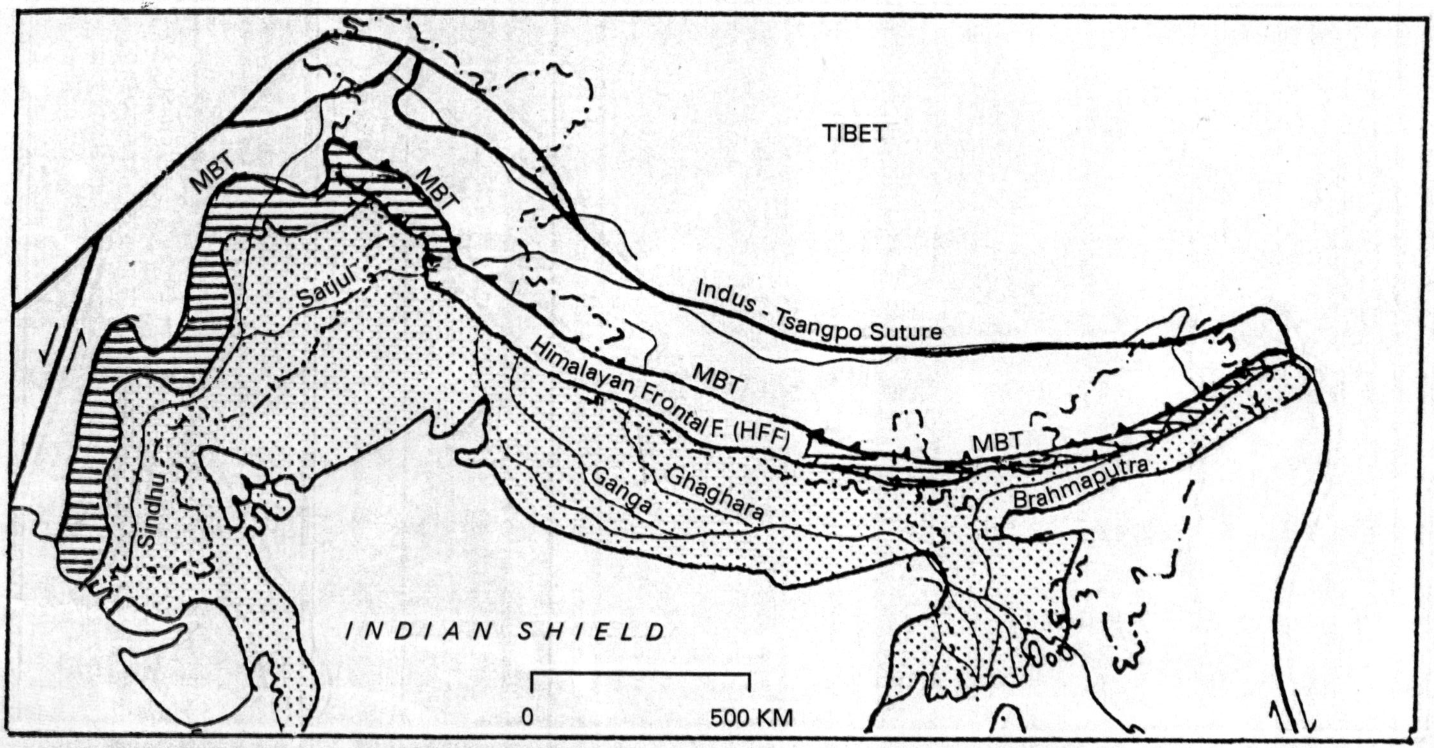


**Figure 8.6** The Himalayan Frontal Fault partitioned the originally large Siwalik Foreland Basin into two unequal parts. The narrow northern part became the Siwalik terrane of rising hills, and the wider southern belt became the subsiding depression. The depression was filled up rapidly with sediments and converted into the Indo-Gangetic Plains.

occurred obliquely along what is known as the Himalayan Frontal Fault (HFF). The northern 25 to 45 km-wide part became the Śiwālik terrane of the rising hills, and the southern 200 to 450 km-wide part became a subsiding depression. This depression was rapidly filled up with sediments and eventually transformed into the Indo-Gangetic Plains.

Concealed under the thick pile of riverine sediments, the Sindhu–Ganga–Brahmaputra Basin is characterized by the very uneven topography of its floor. There are many E–W to ENE–WSW oriented upwarps and depressions, and quite a few transversely oriented (N–S to NNE–SSW and NW–SE) ridges of high relief. These ridges are the northerly extensions of hill ranges of central India, such as the Aravali, the Bundelkhand, the Satpura and the Meghalaya–Mikir massif. Between the ridges are depressions that deepen northwards toward the HFF. The Sharada and the Gandak depressions (Figure 8.6), for example, are 6000 to 7000 m deep and comprise of 1500 to 2000 m-thick columns of Holocene (recent) sediments. The subsidence of the floor of the Ganga Basin must have been due to the flexing down of the crust under the load of the sediments that accumulated briskly.

## The Himālaya Acquires a Morphostructural Design

The 0.85 and 1.6 million-year events represent the diastrophic movements that invested the Himālaya with its unique structural design. It acquired a morphostructural personality of its own, and it became "*Nagādhirāj*" 'the king of mountains'. It may be recalled that the Quaternary culminating events were preceded by five major phases of Himalayan revolution—(i) the convergence of India and Asia, (ii) the collision of continents, (iii) the establishment of drainage of the big rivers of the Himālaya, and the beginning of the formation of the Palaeogene foreland basin, (iv) the emergence of the Himādri, the development of the Lesser Himalayan terrane, and initiation of the Neogene foreland basin, and (v) the emergence of the Śiwālik terrane and formation of the Indo-Gangetic depression.

The Himalayan province was sharply defined in the north by the Indus–Tsangpo Suture against the Karakoram–Tibet region of mainland Asia, and in the south by the Himalayan Frontal Fault against the Indo-Gangetic Plains of the Peninsular India (Figure 8.7). Between the defining faults, the Himalayan province comprises of four lithologically different, structurally distinctive, and topographically contrasted terranes.

The synclinal terrane of the Tethys Himālaya in the north (Figures 8.7 and 8.8) comprises sedimentary-rocks, ranging in age from more than 600

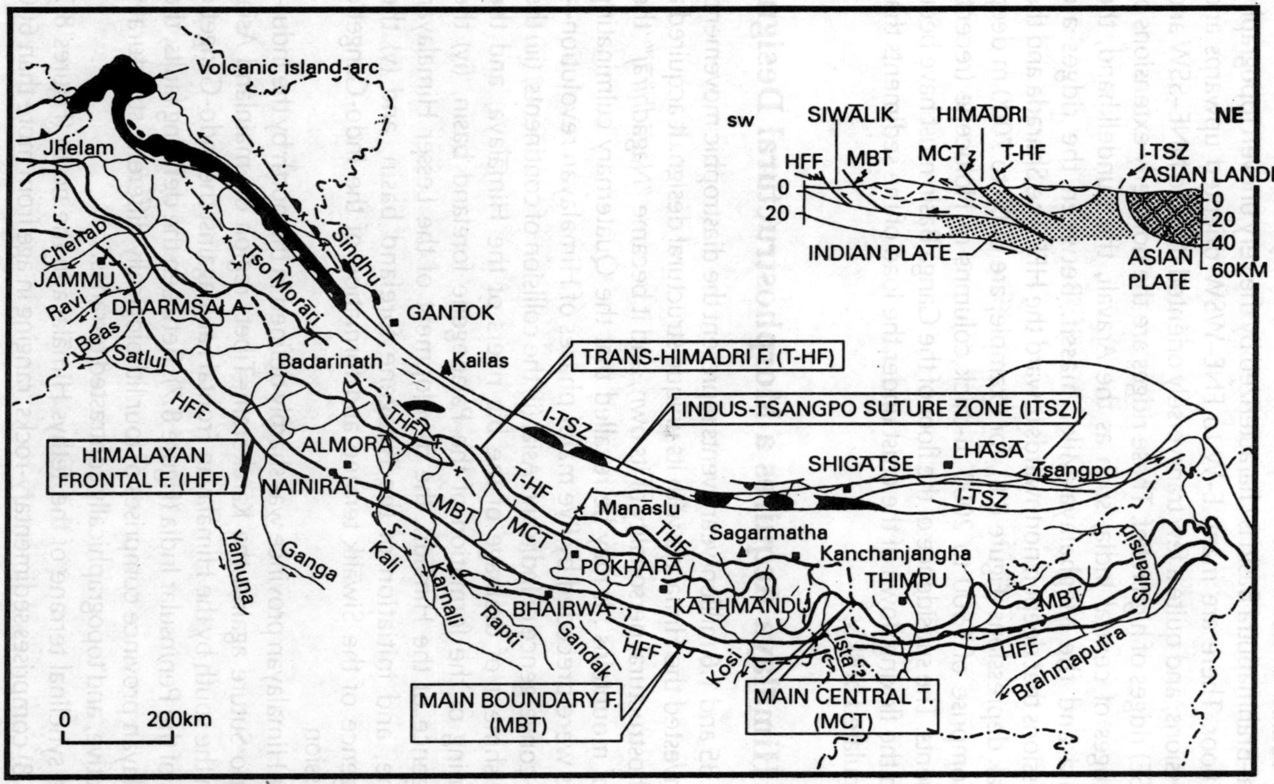


**Figure 8.7** Four boundary thrusts (thick black lines) define the limits of the four structurally distinctive, lithologically different, and topographically contrasted terranes—the Tethys, the Himadri, the Lesser Himalaya, and the Siwalik. The inset depicts the overall structural design of the Himalaya.

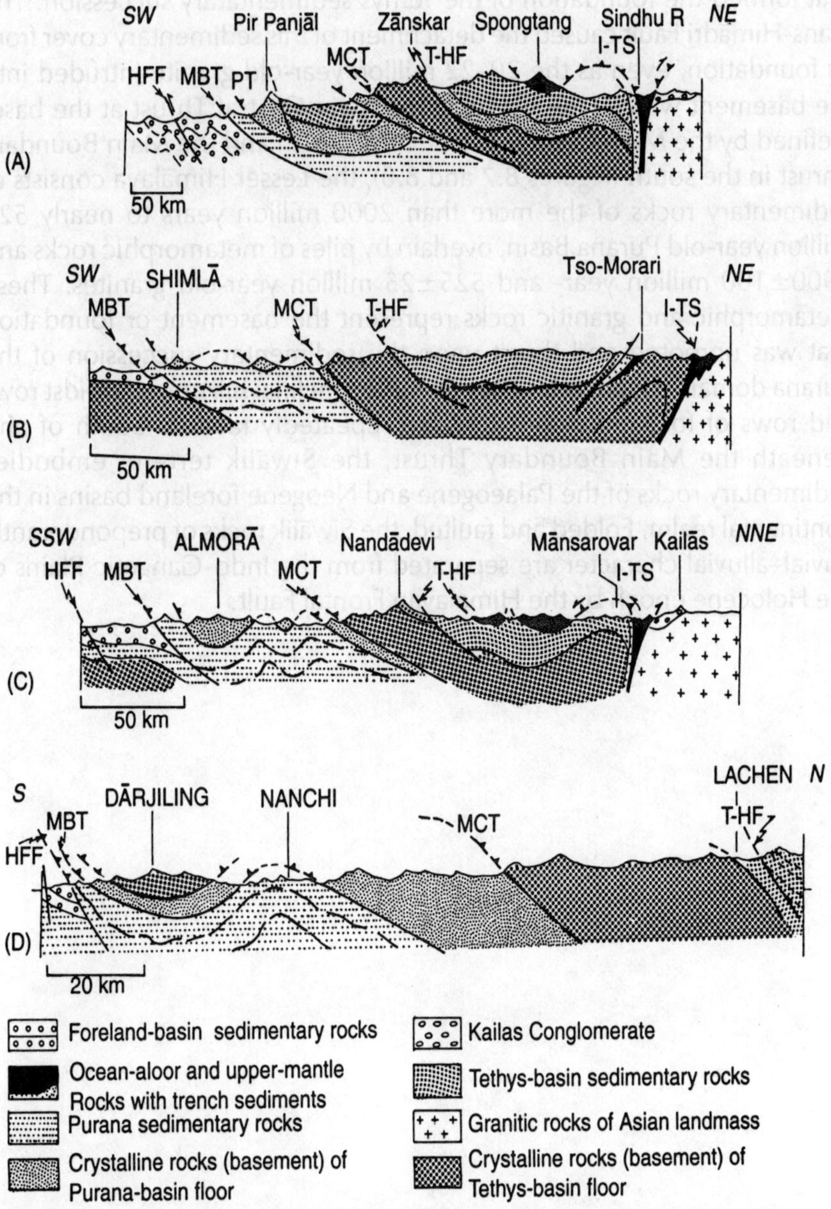


**Figure 8.8** The structural design of the Himalaya is explicit from the cross-sections across the Kashmir, Himachal, Uttaranchal and Darjiling Himalaya. [I-TS — Indus–Tsangpo Suture; T-HF — Trans-Himadri Fault; MCT — Main Central Thrust; PT — Panjal Thrust; MBT — Main Boundary Thrust; HFF — Himalayan Frontal Fault].

million years to a little less than 65 million years, that evolved in the basins of the Palaeotethys and Neotethys Seas. The Himādri terrane is made up of high-grade metamorphic rocks and 500–550 million year-old granite that formed the foundation of the Tethys sedimentatary succession. The Trans-Himādri Fault caused the detachment of this sedimentary cover from its foundation, even as the 20–22 million year-old granite intruded into the basement which rose up along the Main Central Thrust at the base. Defined by the Main Central Thrust in the north and the Main Boundary Thrust in the south (Figures 8.7 and 8.8), the Lesser Himālaya consists of sedimentary rocks of the more than 2000 million years to nearly 525 million year-old Purana Basin, overlain by piles of metamorphic rocks and 1900±100 million year- and 525±25 million year-old granites. These metamorphic and granitic rocks represent the basement or foundation that was uprooted and thrust upon the sedimentary succession of the Purana domain (Figure 8.8). They occur as synclinal hill ranges amidst rows and rows of fold mountains that are repeatedly faulted. South of and beneath the Main Boundary Thrust, the Śiwālik terrane embodies sedimentary rocks of the Palaeogene and Neogene foreland basins in the continental realm. Folded and faulted, the Śiwālik rocks of preponderantly fluvial–alluvial character are separated from the Indo-Gangetic Plains of the Holocene epoch by the Himalayan Frontal Fault.

# 9

# *Evolution of the Indo-Gangetic Plains*

## Development of Duns in the Śiwālik

The strong tectonic movement, that brought about the dismemberment of the Himālayan Foreland Basin into the Śiwālik and Indo-Gangetic domains, profoundly affected the Śiwālik terrane itself. New faults appeared and older ones were reactivated. Wherever anticlinally-folded rocks popped up or the ground moved up or sideways along active faults, there was considerable disruption and interruption of drainage. The uplift of the downstream blocks caused the blockage of streams and rivers, leading to the formation of lakes or marshes in the mountainous Śiwālik (Figure 9.1). Swift-flowing torrents rapidly filled these impoundments. The sediment-fills eventually emerged as flat stretches in the valleys of the hilly Śiwālik domain. These flat stretches are known as *'dūn'*—such as the Pinjor Dun in Haryana, Dehra Dun and Kota Dun in Uttaranchal, and Rapti Dun in Nepal. The *duns* were formed in the Late Pleistocene to Early Holocene time 22 000 to 7 000 yr BP.

## Filling up of the Sindhu-Ganga-Brahmaputra Basin

The floor of the Sindhu–Ganga–Brahmaputra Basin formed south of the active Himālayan Frontal Fault, subsided progressively under the phenomenally growing load of sediments. The subsidence went on concomittant with the piling up of sediments in the floodplain. The

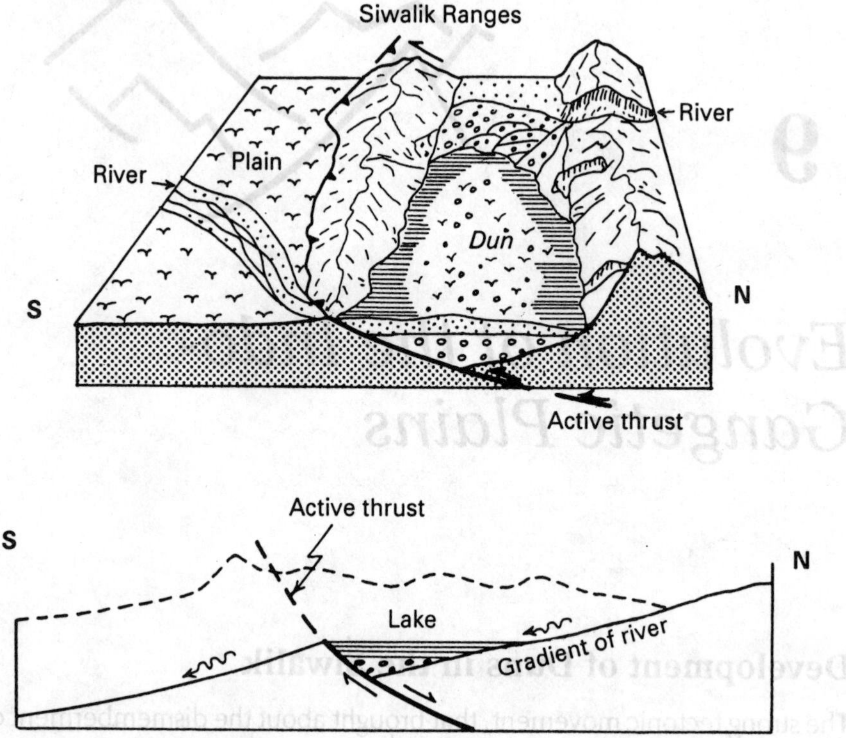


**Figure 9.1** The reactivation of a fault entailing uplift of the downstream block across the path of a river, caused its ponding. Swift filling up of the impoundment gave rise to a flat stretch of sediments called *'dun'* within the hilly terrane.

surface of this floodplain today stands 30 to 100 m above sea level. The thickness of the detrital accumulations, varying from sector to sector, increases progressively northward from less than 500 m in the southern fringe to more than 2500 m along the Gorakhpur–Motihari belt in the north.

The Indo-Gangetic Basin was filled transversely by rivers draining the Himālaya (Figure 9.2) and the uplands of central India. The rivers built fans and lobes (megafans) as they shed their sedimentary loads. The fans coalesced to form a multistoreyed complex of sediments, culminating in the development of the floodplain that is considered to be largest on the surface of the earth. The rapid growth of the megafans pushed the Himālayan rivers southwards, forcing them to flow 200 to 300 km away from the mountain front. Downstream, the increased influx of sediments pushed the lobes and fans to grow seawards in the form of deltas, encroaching upon the realm of the Indian Ocean. Thus evolved the Bengal Basin of Bengal and Bangladesh in the east (Figure 9.3) and the Sindh Basin in the west. The reclamation of the sea through the growth of deltas in Bengal has been going on for over 7000 years. At present, in the terminal reaches of the Ganga River, approximately 1500 billion cubic metres of its sediment load is sequestered within the floodplain, nearly 1970 billion $m^3$ detritus go to build up the Sundarban delta, and about 1500 billion $m^3$ material flows down to the Bay of Bengal.

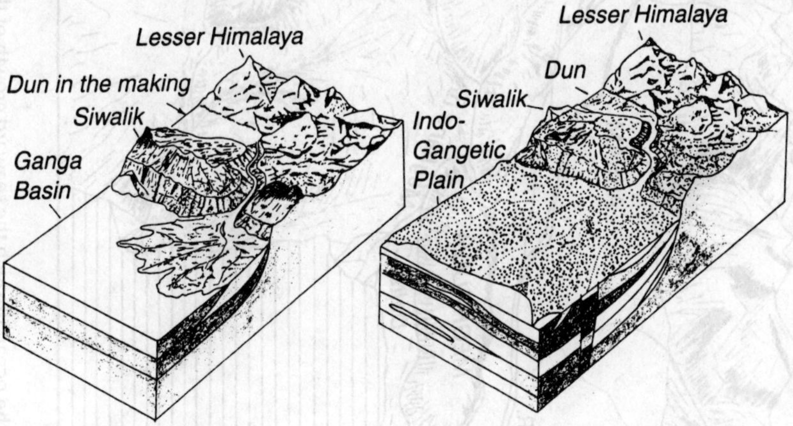


**Figure 9.2** The combined processes of rapid deposition of voluminous sediment and synchronous subsidence of the basin floor gave rise to the Indo-Gangetic Plains.

## Profile of the Plains

The Indo-Gangetic Plains embody the riverine deposits of several generations. In the Uttar Pradesh–Bihar sector, its southern part fringing the ancient hills of Satpura, Vindhya, Bundelkhand and Aravali are made up of the oldest sediments deposited by rivers Son, Ken, Betwa and Chambal that drain the Peninsular Indian Shield. The ongoing uplift of the Vindhya terrane has prompted severe gully erosion, leading to the development of ravine land, particularly in the Chambal–Betwa tract. On the other side in the north, the foothills of the Śiwālik have an apron of gravel and coarse detritus—a chain of coalescing gravel fans and cones described as *Bhabhar* (Figure 9.4). Mountain torrents have been building these fans and cones for quite a few thousand years. This is the youngest unit of the Indo-Gangetic domain.

Between the old and young fringes lie the alluvial plains made up of finer sediments. The larger (comparatively older) part comprises of

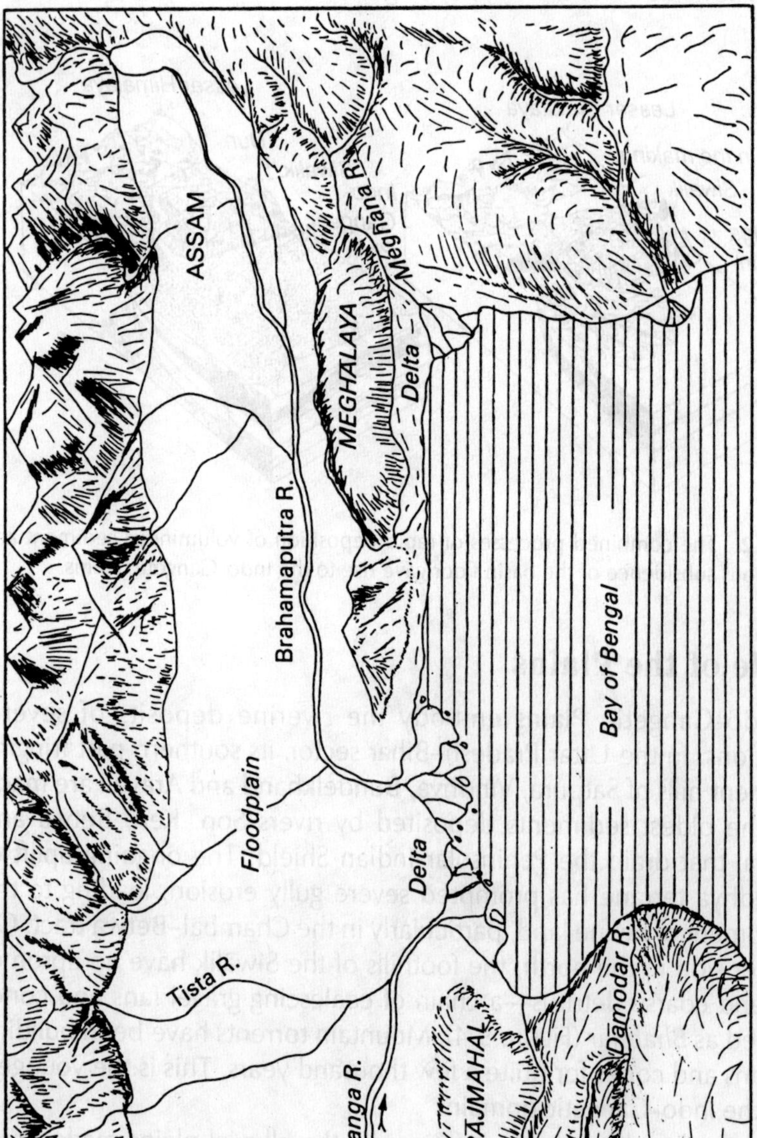


**Figure 9.3** The lands of Bengal and Bangladesh has been won from the ocean through the seaward growth of the deltas that the Himalayan rivers have been building for over 7000 years.

**Figure 9.4** Rivers – like the Gaula and the Kali (Sharada) in southern Uttaranchal – draining the rising mountain, have been shedding their load of sediments and forming fans and megafans. The coalescence of these fans has given rise to the *Bhabhar* apron along the northern periphery of the Indo-Gangetic Plains. *(Photo: NRSA, Hyderabad)*

interfluve (*doab*) deposits called *Bhangar*, forming relatively higher ground including ridges and mounds (Figure 9.5). Concretionary clots of carbonates called *kankar* are characteristic features of the sand, silt and clay deposits of the *Bhangar*. The *kankars* indicate that they were formed in the period 10,000 to 6,000 yr BP, when this part of the subcontinent was afflicted with drier climatic conditions. In the following humid-wet time, meandering rivers deposited silt and clay along with concretionary carbonate in their floodways. These younger deposits are known as *Khadar*. The *Khadar* unit covers large parts of western Uttar Pradesh, Haryana and the Panjab.

Great rivers like the Yamuna, Ganga, Sharada (Ghaghara), Gandak and Kosi continue to build huge lobes of sediments (Figure 9.5), with far-reaching implications for flood hazard and environmental security of the region.

## Climate Fluctuation

In the Himālayan province the Pleistocene glaciers had left behind large volumes of gravel, sand, silt and clay as moraines in their valleys. The

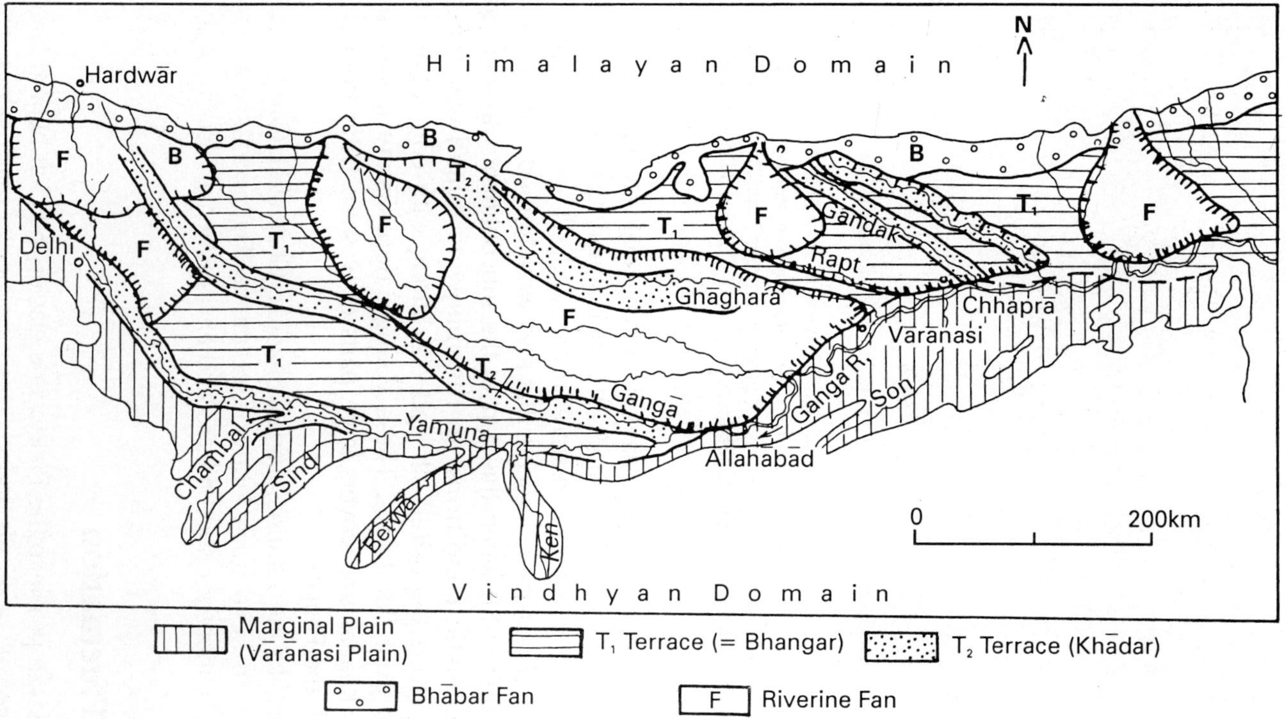


**Figure 9.5** Uttar Pradesh–Bihar part of the Indo-Gangetic Plains embodies riverine deposits of several generations—the oldest unit fringes the upland of the Peninsular Shield in the south and the youngest *Bhabhar* is an apron of gravels along the foothills of the Siwalik. Between the two lie the alluvial *Bhangar* and *Khādar* plains.

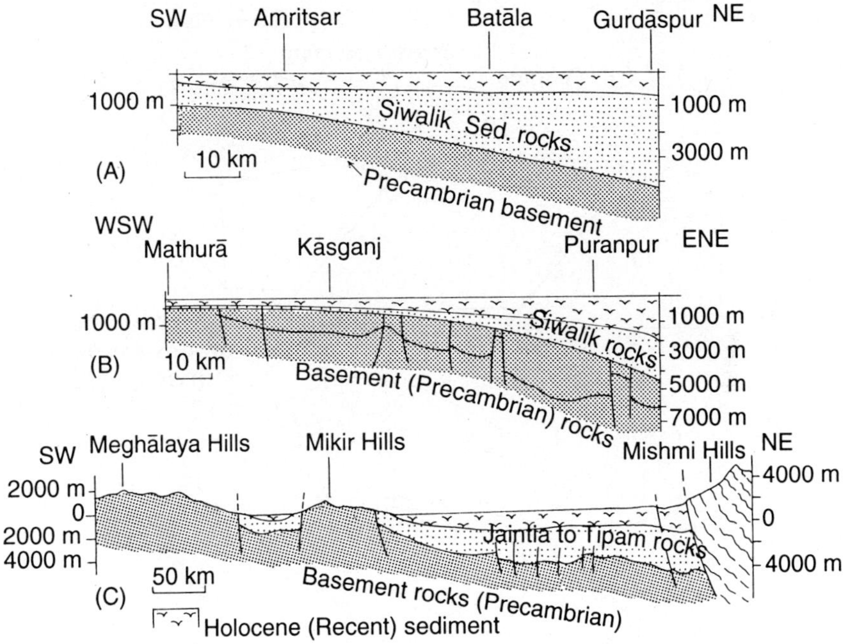


**Figure 9.6** Profiles across the Indo-Gangetic Plains bring out the variable nature of the floor of the basin and the progressively northward-increasing thickness of the Holocene (Recent) sediments.

cold-dry wind that started blowing after the retreat of the glaciers carried away the finer sediments and spread them far and wide as loess. In the Kashmir valley, for example, the loess formed a blanket of silt on the Karewa lake deposit. Within this 25 m-thick mantle of loess , occur many layers of fossilized soil – palaeosols – rich in organic matter, including lignite in some places (Figure 9.6). The carbonaceous matter of the palaeosols represents short, warm-wet spells within the otherwise long, dry-cold climate. In other words, the climate fluctuated between dry-cold and warm-wet in the nearly 200,000 year-long period that followed the end of the glaciation.

Since the Himālayan province continued to be a hotbed of tectonic ferment, there were many events of pronounced reactivation of older faults. Movements along these faults 40 000 to 60 000 years ago brought about river impoundment, wherever the downstream blocks were uplifted across the drainage, in several parts of the mountain province (Figure 9.7). Some of the lakes thus formed in the upper reaches of streams persist to the present, such as the Rara Lake in western Nepal, the Naini*Tal* and Bhim*Tal* in southcentral Kumaun (Figure 9.8A), and the *Tso*kar in southeastern Ladakh. Others are represented by accumulations of brown and black carbonaceous clay, such as the Wadda palaeolake in eastern

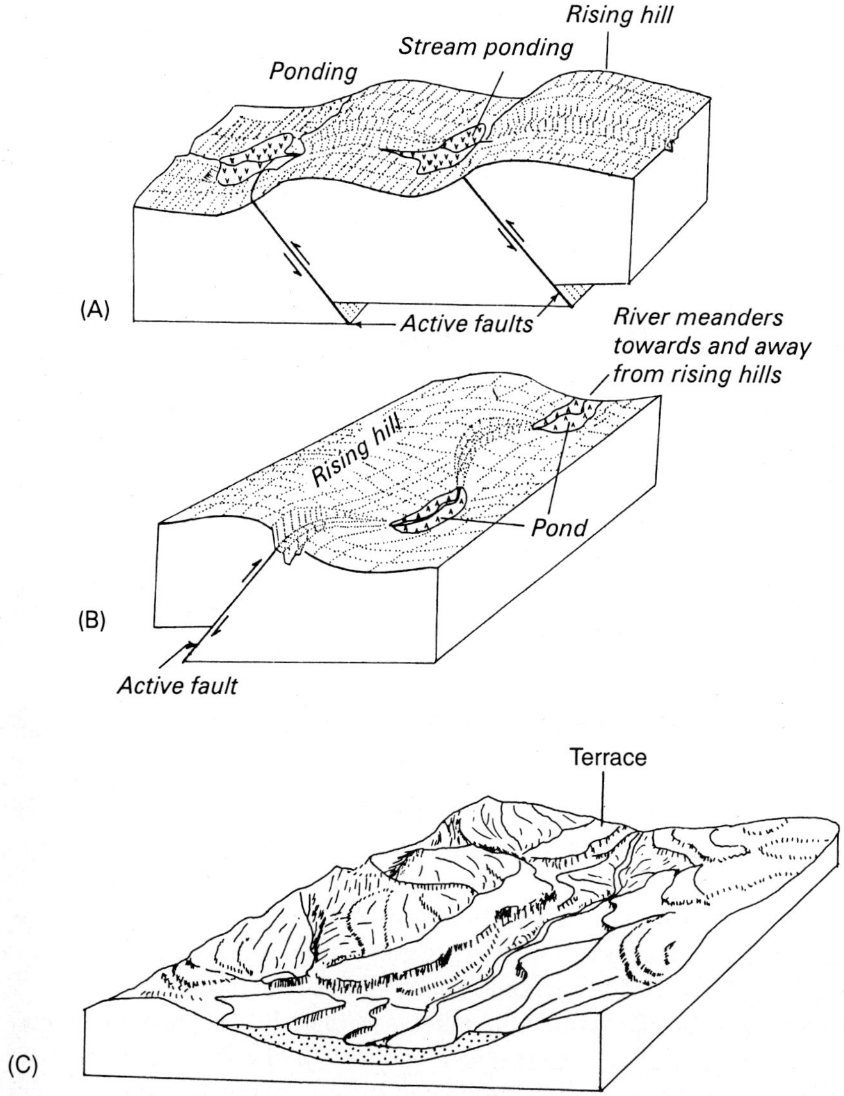


**Figure 9.7** The uplift or subsidence of the ground, following the reactivation of faults, caused the ponding of streams that crossed or followed them. (Based on a diagram from the 1994 paper by G King and others). C) shows the remains, in the form of paired terraces, of the lake that disappeared later due to draining out of the impounded water.

Kumaun, the Lamayuru palaeolake in south-central Ladakh (Figure 9.8B), and the Skardu palaeolake in the Karakoram beyond the Sindhu River.

Pollens and spores embedded in the sediments of the lakes, mentioned above, throw light on the climatic conditions of the past 50,000–60,000 years. The prolonged cold-dry period ended about 11,000 yr BP, marking

(A)

(B)

**Figure 9.8** (A) Naini Lake (Naini*Tal*) originated nearly 40,000 years ago when the eastern block (left side) rose up along an active fault, causing the blockage of a petty stream. *(Photo: Anup Sah)* (B) Lamayuru palaeolake in southwestern Ladakh, formed more than 35,000 years BP, is represented by a nearly 200 m-thick succession of sediments characterized by remains of plants and animals, and also conspicuous rosettes of gypsum. This evaporite mineral is indicative of the prevalence of warm-dry climate once upon a time in the lifespan of the lake that ended between 1,000 to 500 yr BP. *(Photo: TN Bagati)*

**Holocene Climate Changes**

| Area (Testimony of) | Climate: Warm-Wet (yr BP) | Climate: Hot-dry (yr BP) |
|---|---|---|
| Arabian Sea | 10,500– 5,000 | 3,500 |
| Western coast (terrestrial detritus) | 10,500–10,000 | |
| Karwar coast (land-derived pollens) | 10,500– 5,000 | 3,500 |
| Nilgiri Hills ($\delta^{13}$ C values of peats) | 9,000– 8,000 | 5,000–2,000 |
| Didwana-Lunkaransar lakes in Rajasthan (pollen profiles) | 10,800– 4,500 | 4,000–Present |
| Gangotri Glacier (pollen in sediments) | 6,500– 4,000 | 3,500–3,000 |
| Tsokar Lake, Ladakh (*Juniperus* flora) | 10,000 | |
| BangongCo Lake, Aksai Chin (pollens) | 6,000 (max.) | 4,000–3,000 |
| Southeastern Tibet (pollens in lakes) | 7,500– 3,000 | 3,000–1,500 |
| Bhimtal Basin, South-central Kumaun (pollens) | | 3,550±120 |
| Rara Lake, West Nepal (pollens) | | 4,500 |
| Karewa Lake, Kashmir (pollens) | | 5,000 |

**Figure 9.9** Fluctuation in the climatic conditions during the Late Pleistocene and Early Holocene periods, as recorded by pollens recovered from lake sediments in different parts of the Indian subcontinent.

the end of the Pleistocene and the beginning of the Holocene epoch. Then followed a period of rainfall under a warm climate from 11,000 to 3,500 yr BP. It was very hot and dry during the period 3500 to 2000 yr BP all over the subcontinent, in the Indian Ocean, and in the Tibetan Plateau (Figure 9.9).

## Past Animal Life in Indo-Gangetic Plains

The fossil remains retrieved from the Indo-Gangetic sediments indicate that the animal life was not very different from what it is at present. The animals included, among others, elephants, buffaloes and cows. Some species of these animals have become extinct, such as the hippos *(Tetraprotodon palaeindicus* and *Hexaprotodon)* the elephant *(Elephas namadicus)* the primate *(Semnopithecus)* and the cow *(Bos namadicus)*. It must have probably been the drastic changes in climate that led to their extinction.

Skeletal remains of human-like primates and of early man have not been found so far in the Indo-Gangetic domain, despite the fact that the ape *(Sivapithecus)* lived in the woodlands of the Lower Śiwālik time in the Harytalyangar–Jammu belt and stone implements have been found from sediments as old as 540,000 yr BP. This has led to the speculation that man evolved outside the Indian subcontinent and came in here only as the evolved species—*Homo sapiens*.

## Appearance of Man

Judging from the finds of a variety of stone implements, it appears that tool-making creatures have been living in India ever since >540,000 yr BP. The discovery of stone artefacts in a number of "duns" and lake basins, such as Pinjor, Suketi, Rapti and Karewa, indicate that the tool-making creatures had found a foothold in the Śiwālik terrane, and that they preferred living alongside lakes and rivers. The earliest were the people of the Palaeolithic culture who lived in extreme northwestern Kashmir, in western Rajasthan, Gujarat, and adjoining parts of Madhya Pradesh and Maharasthra. However, fossils of the human skeleton appear only after 45,000 to 50,000 yr BP.

It is speculated that the people who speak Austric languages migrated from Africa and entered India, via northeastern India, about 50,000 years ago and settled down in eastern and central India (Figure 9.10). Speakers of the Dravidian languages – who knew the art of cultivating wheat and domesticating cattle – came from West Asia between 8,000 to 10,000 yr BP. They occupied lands in western and southern India. People of the Sino–Tibetan stock, who cultivated rice and reared buffaloes, came from Southeast Asia between 8,000 and 10,000 yr BP and settled down in the northern territories of the Himālaya and the hills along the border with Myanmar. The people of the Indo-European stock came in several waves of migration between 3500 and 6000 yr BP from Central Asia. They rode horses and used iron technology. These immigrants made northern India, including the Himālayan province, their home.

Sometime between 10,000 and 11,000 yr BP, the Indian subcontinent came once again under the sway of heavy rains, and the climate became progressively wetter but warmer (Figure 9.9). The land was clothed in a green carpet of vegetation and the valleys in the Himālaya were verdant. Naturally, the prehistoric people were attracted to this land of rivers and plentiful food. Going by the evidence provided by spores and pollens recovered from the sediments of the Lunkaransar and Didwana lakes in western Rajasthan, the rainfall had become very heavy in the period 4000 to 6500 yr BP. Carbon dating shows that the Lunkaran lake was brimming with water from 6300 yr BP to 4800 yr BP. and that humans were in occupation of the lake basin about 4230±55 yr BP. That was indeed the time when the whole of northwestern India, including western Rajasthan and all the terranes of the Himālayan province, was dotted with the settlements of prehistoric people (Figure 9.11).

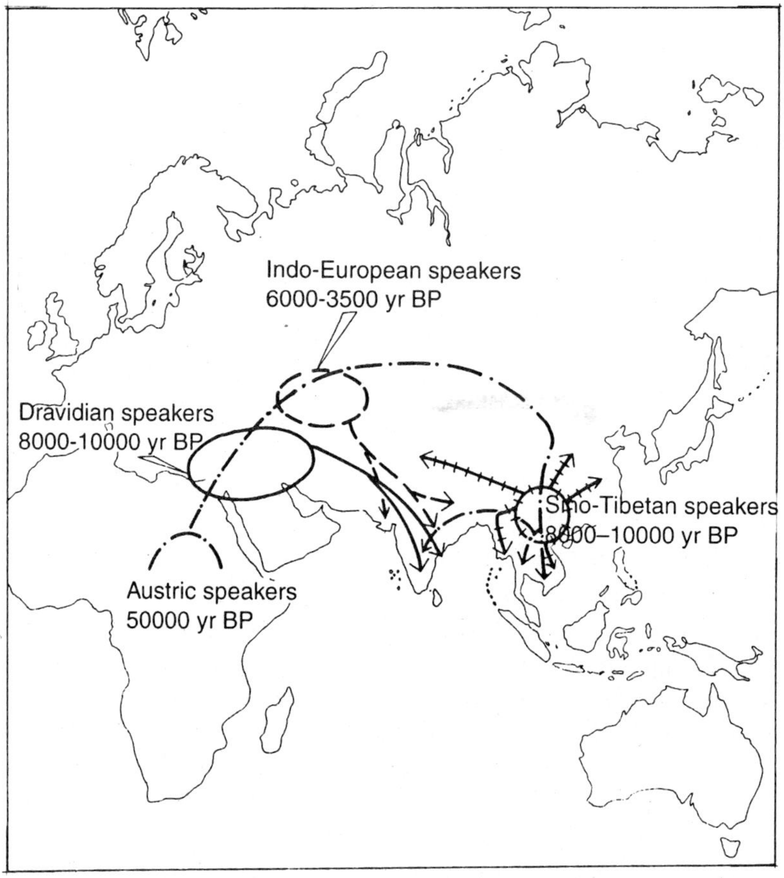


**Figure 9.10** The peoples who inhabit the land of India today came in successive waves of migration from different centres of evolutionary expansion. (A speculative diagram based on the 1998 paper by Madhav Gadgil and others.)

# Human Settlements

Significantly, among the vegetal remains recovered from the sediments of the Lunkaransar and Sambhar lakes are cutigens, pollens and charcoal of the stubbles of cereal plants. The charcoal is dated 8000 to 9400 yr BP. Clearly, the people who lived around these lakes had started harvesting food-crops. The domestication of plants, like the domestication of animals, marks a crucial development—the beginning of a new mode of life. The hunter-gatherer had become a producer-herder. His life became relatively sedentary. This led eventually to the emergence of urban culture—the culture of the Harappan people who lived in their urban centres in the floodplains of the Sindhu and Saraswati Rivers in the early to middle Holocene times.

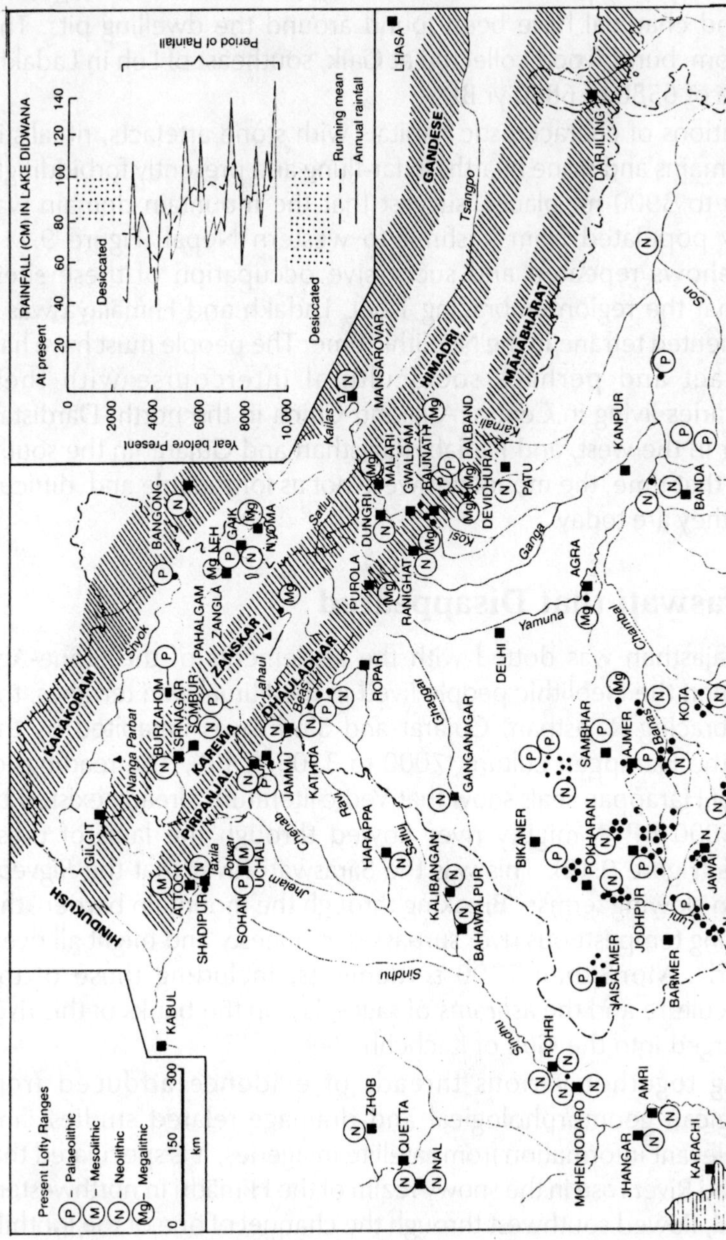


**Figure 9.11** Stone-Age (Prehistoric) settlements (shown by solid circles) in northwestern India.

The Stone-Age people lived around lakes, generally in the intermontane flat stretches of land—such as the Karewa basin in Kashmir, the Pinjor Dun in southeastern Himachal Pradesh, and the Rapti Dun in Nepal (Figure 9.11). The Burzahom people, near Srinagar in Kashmir, lived in oval-shaped pits with an overhead wooden superstructure of sorts. They buried their dead in a crouching posture along with their pet animals, using ash, lime and potsherds as burial material. Burnt bones, stone artefacts and charcoal have been found around the dwelling pits. The charcoal from burnt wood collected at Gaik, southeast of Leh in Ladakh, gives a date of 6580 to 6840 yr BP.

The locations of characteristic habitats with stone artefacts, ritualistic funerary remains and stone hearths in far-flung and presently forbiddingly high (1600 to 3900 m) places, suggest that the mountain domain was extensively populated from Kashmir to western Nepal (Figure 9.11). Evidence shows repeated and successive occupation of these sites, implying that the region embracing Tibet, Ladakh and Himālaya was a much frequented terrane in the Neolithic time. The people must have had easy contact and perhaps sociocultural intercourse with their contemporaries living in Central Asia and China in the north, Dardistan and Potwar in the west, and Panjab, Rajasthan and Gujarat in the south. Possibly, at that time, the mountains were not as formidable and difficult to cross as they are today.

## The Saraswati that Disappeared

Western Rajasthan was dotted with the settlements of the Stone-Age people. When the Neolithic people lived in the Himālayan terranes, the region embracing Rajasthan, Gujarat and Sindh was inhabited by the people of the Harappan culture (7000 to 3300 yr BP). The readings of hundreds of Harappan seals show that Vedic literature already existed by the year 5000 BP. A mighty river flowed through the land of these settlements (Figure 9.12). This was the Saraswati River that the *Rigveda* describes in glowing terms: "Breaking through the mountain barrier" this "swift-flowing tempestuous river surpasses in majesty and might all rivers of the land". More than 1200 settlements, including those of the Harappan culture and the *ashrams* of sages, lay on the banks of this river that discharged into the Gulf of Kachchh.

Weaving together various threads of evidence adduced from archaeological, geomorphological and drainage-related studies, and gleaning relevant information from satellite imageries, it is speculated that the Saraswati River rose in the snowy realm of the Himādri in northwestern Uttaranchal, flowed southwest through the channel of one of the foothills tributaries of the present-day Ghaggar River and met the then southeast-flowing Shatadru (Satluj) of that time at Shatrana about 15 km south of Patiala. At the confluence, the channel was 6 to 8 km wide, implying the great discharge of the River Saraswati. The Ghaggar is known as the Hakra in its middle reaches and as the Nara in the lower reaches. Significantly, the water recovered in the middle reaches from tubewells deeper than 60 m was found to be 22 000 to 6 000 years old, and shallow-well water was found to be 5000 to 1800 years old. The age of the water (indicated by carbon dating) increased downstream from Kishangarh. Since the tritium value is negligible, these waters could not have been fed through contemporary recharge. The deeper – and older – water must be attributed to the ancient river that flowed in the Late Quaternary time earlier than 5000 yr BP.

Tectonic movement, including the uplift of the NE–SW trending fault-delimited blocks of the Aravali Range, must have caused the deflection of the headwaters of the Yamuna and Shatadru, leading to the disappearance of this mighty river. The eastern branch – what is now the Yamuna – deviated southwards around 3,700 yr BP, flowed through the channel of a tributary of the Chambal, and joined the Ganga at Triveni or Allahabad

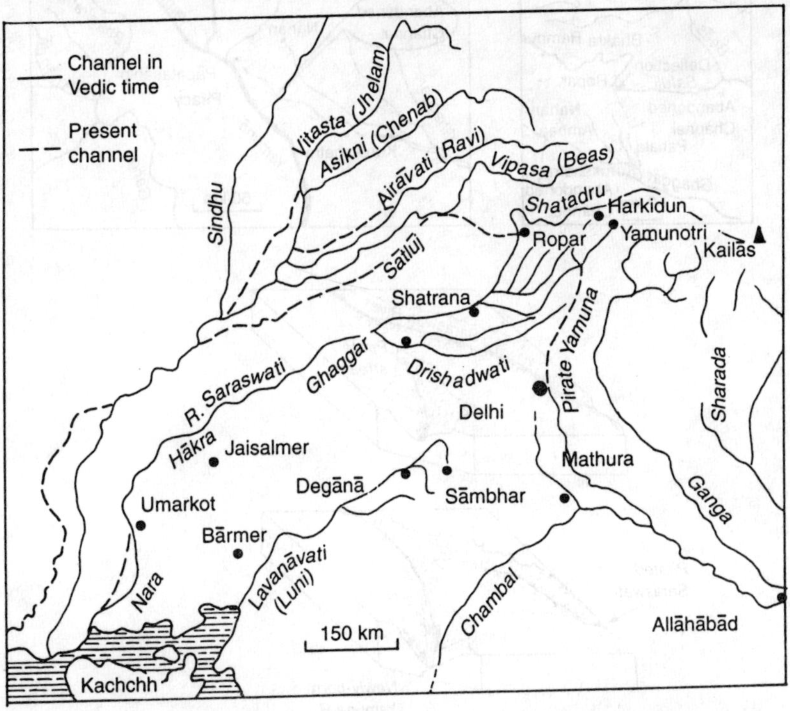


**Figure 9.12** The Saraswati River that was formed by the confluence of the Yamuna of that time and the Shatadru (Satluj), was a mighty river during the Vedic time. It is now represented by the dry channels of the Ghaggar, the Hakra and the Nara.

(Figure 9.13). The consequent dwindling of the river discharge propelled the migration of the Late Harappan (3900–3300 yr BP) people upstream from the Ganganagar–Bahawalpur area to the upper reaches of the Śiwālik. This is evident from a dramatic increase of the Late Harappan settlements in the Śiwālik belt in southeastern Himachal Pradesh and the adjoining Haryana and Uttar Pradesh. As a matter of fact, this region became populated for the first time.

Later, during the time Gautam Buddha lived in the east, about 2600 yr BP the Shatadru River also betrayed the Saraswati as it abruptly swerved westward to join the Beas of the Sindhu River system. Deprived of the waters of these two major rivers, the Saraswati became a dry channel. The collapse of the Harappan civilization seems to be wholly due to the disappearance of the Saraswati and its associated rivers.

As tectonic activity continued to afflict the region, there were frequent disruptions, including changes of the river course. It seems that some

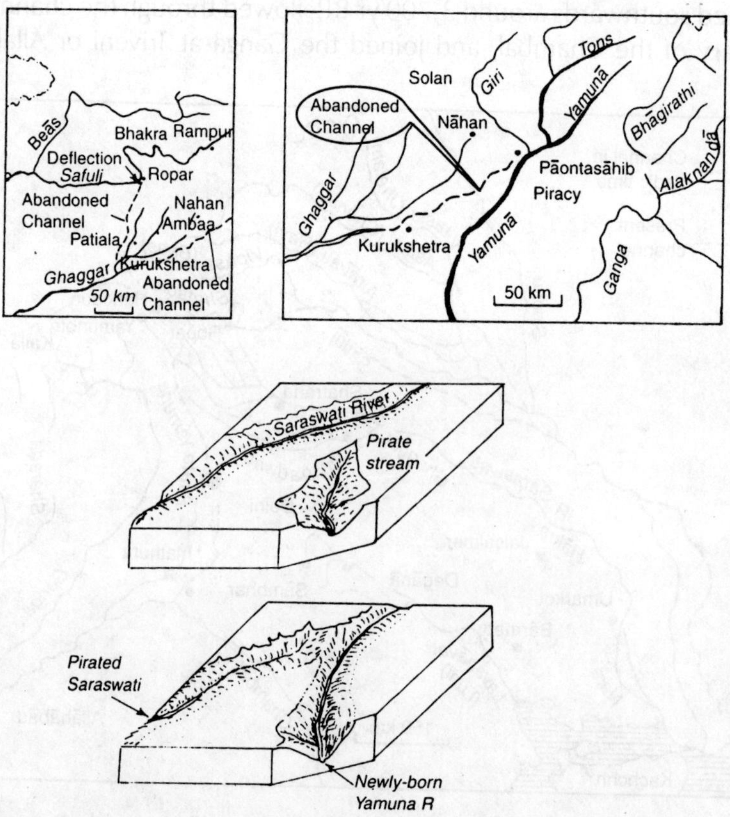


**Figure 9.13** Deflection of the headwaters of the Saraswati River – the Yamuna as well as the Shatadru later – caused the disappearance of the once-mighty river.

branch of the constantly-shifting Shatadru continued to flow into the Saraswati. This is evident from the existence of settlements of the Painted Greyware Culture in the Ganganagar area in the first millenium BC. The old Saraswati River remained a dry channel for several centuries, but eventually some water found its way again into it in the early centuries of the Christian era, as borne out by the Rangmahal settlements in the Ganganagar–Bahawalpur segment. The combined Beas-Satluj – called Biyeh – flowed from 600 BC until about 1100 AD and took the course of the Hakra-Nara to discharge into the Gulf of Kachchh. Sometime after the Biyeh went west to join the Sindhu, the Hakra-Nara River became absolutely dry. This development in the historical time forced a fresh wave of migration of the people to greener pastures elsewhere.

# 10

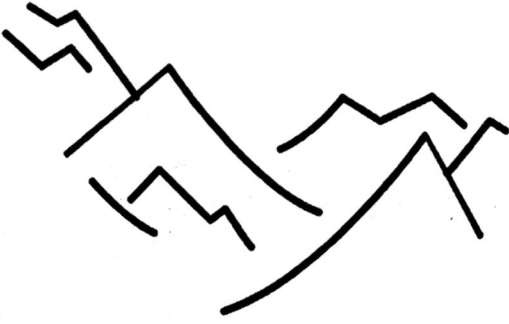

# *Tectonic Tumult Goes on*

The northward movement of the Indian plate is going on at the rate of approximately 5.5 cm/yr (Figure 10.1). Sandwiched between the continents of Asia and India, the Himālaya is under very strong, persistent compression. Therefore, there is a steady build up of strain in its giant frame, particularly in the zones of faults that demarcate the boundaries of the four geological–physiographic terranes (Figures 8.7 and 8.8). In other words, elastic strain is progressively accumulating in the zones of the Himalayan Frontal Fault (HFF), the Main Boundary Thrust (MBT), the Main Central Thrust (MCT) and the Trans-Himādri Fault (T-HF). The relaxation of strain is manifest in the rupturing of the crust – or reactivation of the old rupture planes – in the twitching and shuddering of the mountains (Figure 10.2), in the uplift of the peaks, and in the dislocation of rock masses and their advance over younger sediments.

## Rupturing and Shuddering of Ground

When the accumulated strain exceeds the breaking point of the crust, the ground ruptures. The snapping of the crust and the attendant slipping of rock masses on the rupture planes (faults) give rise to earthquakes. The pattern of the distribution of epicentres – the ground areas above the source of tremors – are related to movements on the active faults, particularly the MBT and its subsurface extension beneath the pile of Lesser Himalayan rocks (Figure 10.2). It is a shallow-dipping plane that separates the pile of Himalayan rocks from its foundation or basement. Tear faults cutting across the mountain domains are also responsible for earthquakes in many a sector.

**Figure 10.1** The northward-moving Indian plate (at the rate of 5.5 cm/yr) is strongly and persistently pressing against the Himalaya.

The belt of frequent earthquakes of magnitudes 5 to 6 is confined to the Lesser Himālaya, which is approximately 50 km south of the MCT that defines the southern boundary of the Himādri terrane (Firure 10.3). The hypocentres lie at depths of 20 to 25 km – and locally up to 30 km – below the surface. The epicentres of small earthquakes are also clustered in this belt and occur in dense concentrations in some pockets. An area of extraordinarily high seismicity is seen in northwestern Nepal and the adjoining northeastern Kumaun region (Figure 10.3). The present-day intense microseismicity in this region suggests an ongoing build up of strain in the deeper zone of active dislocation or detachment that the epicentral distribution defines. Then, there are several pockets of high seismicity indicated by the close clustering of epicentres in the regions cut by transversely-oriented tear faults—in eastern Arunachal, Bhutan, central Nepal, northwestern Nepal and adjoining northeastern Uttaranchal, southeastern Ladakh, and Kashmir. Apparently, the reactivation of the these tear faults is responsible for the more-than-normal seismicity in these areas.

Four times in the last century, the Himalayan arc was ruptured by great earthquakes of magnitude greater than 8—in 1904 in the Kangra sector (M 8.4), in 1934 in southcentral Nepal (M 8.1), in 1897 in the northern fringe of the Meghalaya massif (M 8.7), and in 1950 in northeastern Arunachal Pradesh (M 8.7). These great earthquakes relaxed sizeable proportions of the strain that had accumulated over the earlier few hundred years. The strong earthquake of 1833 in central Nepal (between

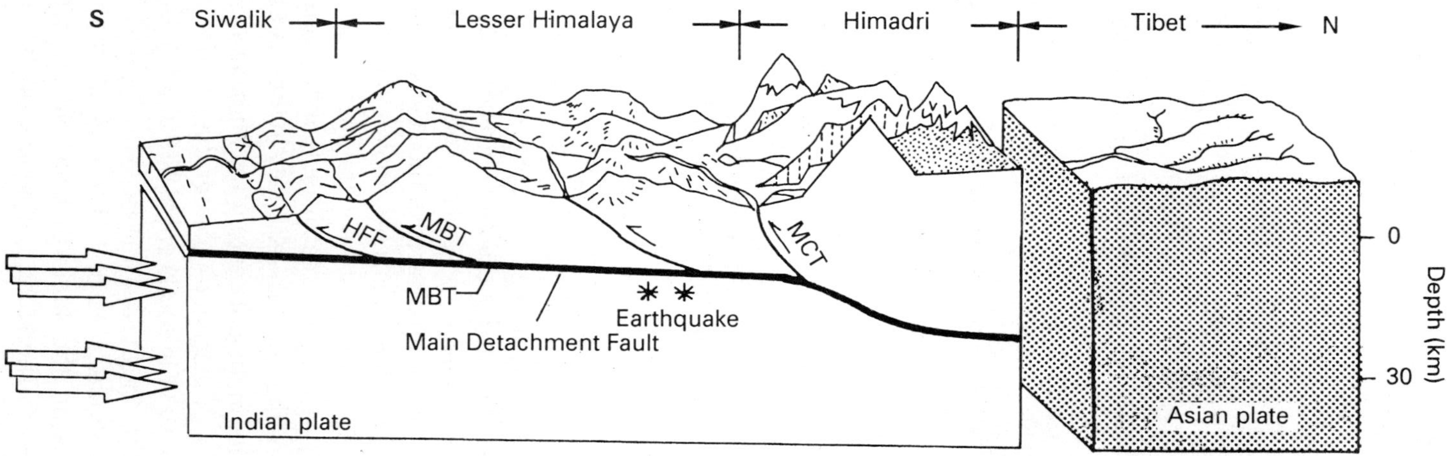


**Figure 10.2** Earthquakes are related to movements along the faults that constitute the boundaries of the four Himalayan terranes—HFF, MBT, MCT and T-HF. Hypocentres (foci) located deep down close to the Main Detachment Plane testify to the reactivation of these faults. (Modified after a diagram in the 1994 paper by M Jackson and R Bilham).

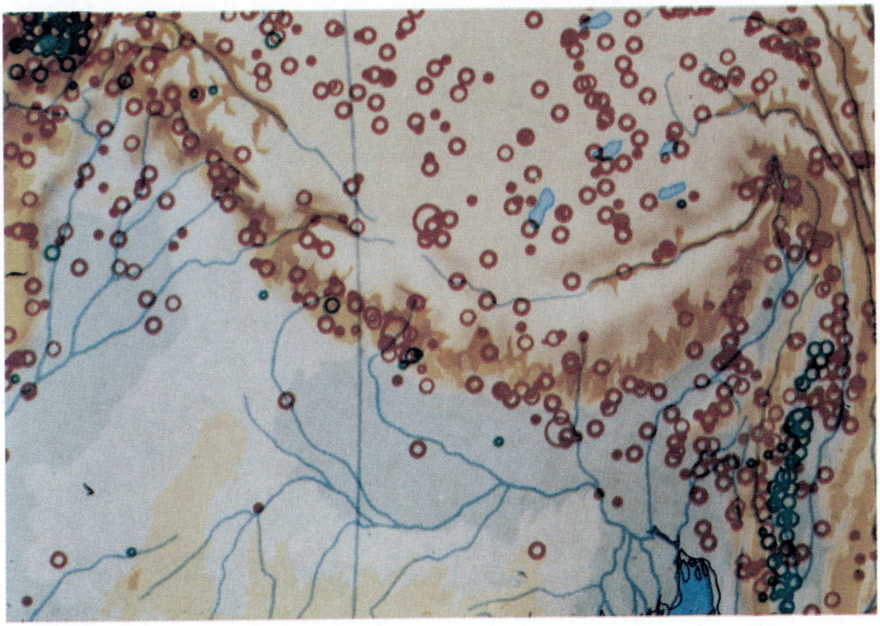

**Figure 10.3** The distribution of epicentres indicates snapping of the crust and attendant dislocation or movement of rockmasses along active faults. It may be noted that the four great earthquakes of the last century ruptured relatively small segments (each 250–300 km) of the mountain arc. Large parts of the arc remain unruptured and, therefore, vulnerable to future earthquakes of larger magnitudes.

M 7 to 7.9) and the 1803 earthquake in Garhwal could not have relaxed the stored strain sufficiently to obviate recurrence of major earthquakes in the central sector embracing western Nepal, Uttaranchal, and eastern Himachal. The amount of creep recorded in central Nepal is too small to ensure aseismic (without earthquakes) accommodation of the convergence of India towards Asia. In other words, in this long central sector of the Himalayan arc, the movements that generate earthquakes have got stuck. The faults are thus locked in the region embracing eastern Himachal, the whole of Uttaranchal, and western Nepal. Seismologists describe this sector as a seismic gap. The unlocking of these faults of the seismic gap, in the event of the strain build-up exceeding the critical point, would cause great earthquakes, the magnitude of which would be over 8.

## Overriding Rock Masses

The ongoing movements on the active faults have brought older Himalayan rocks to settle upon young sediments deposited in lakes, on river banks, or in the foothills. The very old metamorphic rocks of the Mishmi mountain

(A)

(B)

**Figure 10.4** (A) The recent movement along the Main Boundary Thrust is evident from the overriding of very young sediments by very old rocks in the Ladhiya valley near the India–Nepal Border. (B) The reactivation of the HFF in the Recent Time is manifest in the formation of a high scarp, overlooking the Indo-Gangetic Plains, due to faulting up by 60 to 90 m (along the HFF) of the Dun Gravel and its 6° tilt in south-central Kumaun in Uttaranchal.

in eastern Arunachal, for example, override the subrecent gravels and sands of the Lohit River; and the Precambrian rocks of the Darjiling Hills have advanced 15–20 km southwards to rest upon the upwarped riverine sediments of the Tista valley. In southeastern Uttaranchal, the MBT has brought a pile of sheared–shattered Precambrian rocks onto the recent gravel of the Ladhiya River, just before it joins the Kali River (Figure 10.4A). In the Balia valley near Nainital, subrecent river terraces along with younger landslide debris cone have been cut and uplifted to 30–40 m by the reactivated MBT. The Dun Gravel of the Late Pleistocene to Early Holocene age is not only tilted 6° to 15° northwards, but faulted up to give rise to impoundments in the interior of the Śiwālik terrain in the area of the Jim Corbett National Park. The same Dun Gravel has been uplifted along the HFF, as testified by the 60–90 m-high scarp overlooking the plains in southwestern Kumaun (Figure 10.4B). In the place where the Kalagarh Dam is located, a fluvial terrace characterized by tons of pottery pieces of the historical Kushan period (2nd–4th century AD) was lifted up 67 m in the last 1600 to 1800 years.

In the Panjab Śiwālik domain, the recent–subrecent stream sediments have been thrust over and now lie under a pile of older rocks of the Śiwālik. In Jammu, the Precambrian limestone of the Vaishnodevi mountain has advanced southwards, trampling and crushing the subrecent scree at the foothills. Similar developments are seen along the MCT and T-HF in many places throughout the Himālaya. In a number of places, the uplift of the downstream footwall block has given rise to stream impoundments. The lakes were rapidly filled up by gravel and sand, creating flat gravel plains within the steep valleys, in the formidable terrain of the high, rugged mountains.

## Rising Mountains

The movements along active faults consist of both slow continual creep and episodic uplift occurring suddenly in spurts ,with bursts of energy in the form of earthquakes. Between the active faults, compressed folds that form mountain ranges are also being squeezed up and therefore gaining height. The ongoing rise or uplift of the mountains is being measured through geodetic surveys by relevelling and global positioning system studies (GPS). The relevelling done in some sectors of the Himālaya demonstrates that the rate of uplift varies from terrane to terrane, and sector to sector. It is 0.8 to 1.0 mm/yr in the Śiwālik of the Dehradun sector; 3 to 5 mm/yr in the Lesser Himālaya in central Nepal; 5 to 9 mm/yr in the Himādri in north-central Nepal (Figure 10.5A); 3 to 5 mm/yr in the Namcha Barwa knot in the northeastern extremity; and above 9 mm/yr in the Nanga Parbat–Haramosh massif in northwestern Kashmir.

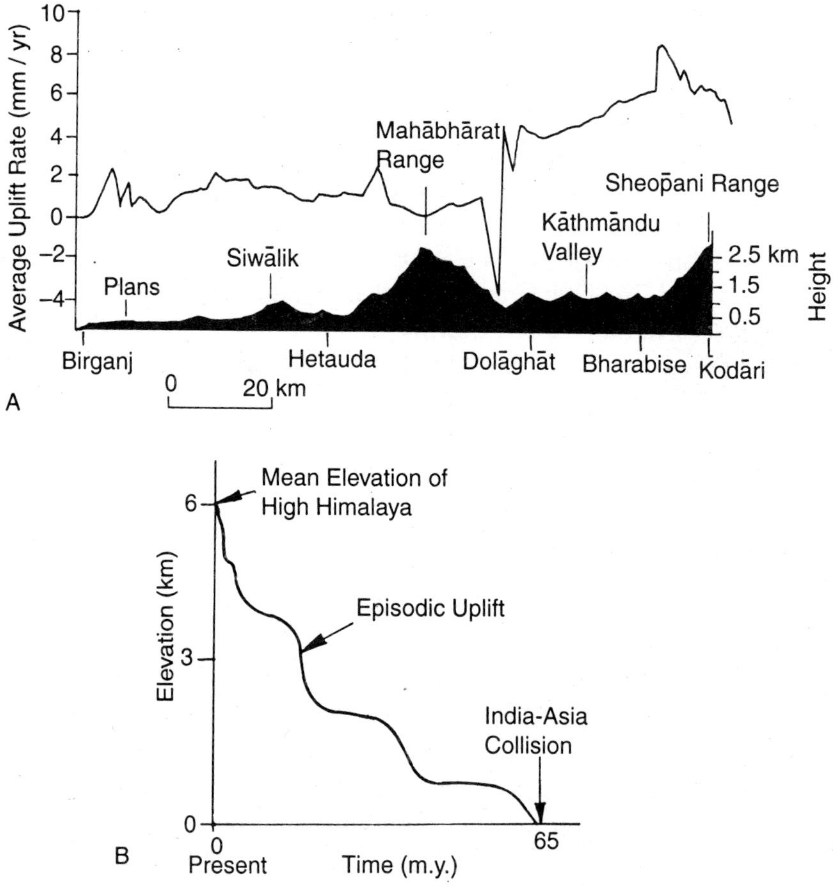


**Figure 10.5** (A) Relevelling done over the period 1977–1990 in the central sector of the Nepal Himalaya showed that the Himadri is rising at the rate of 5 to 9 mm/yr, while the Lesser Himalaya was uplifted at the rate that varied from 3 to 5 mm/yr.
(B) Over a long geological period, the progressively increasing rate of uplift of the Himalaya has been episodic in trend.

The Himālaya, on the whole, is thus rising at an abnormally rapid rate. The uplift rate, over the geological period, has been progressively accelerating. The pattern of uplift on the whole is episodic, occurring suddenly in rapid spurts, as mentioned above (Figure 10.5B).

## Accelerating Erosion

The continuing uplift of the mountain ranges has prompted faster denudation, and attendant landform modifications. The quickening of the pace of erosion and the incidences of landslides are particularly pronounced in the zones of repeated faulting and earthquakes. The mountain slopes in these vulnerable belts have become unstable and succumb easily to rains and human tampering with the ecological balance. The denudation rate in the Sindhu valley in northwestern Kashmir is of the order of 2 to 12 mm/year; in south-central Kumaun about 1.73 mm/yr; and in southeastern Nepal above 1.75 mm/yr. Denudation implies loss of the nutrient-rich soils that support vegetation. Accelerated erosion has therefore had a considerable impact on the integrity of the floral life in the vulnerable Lesser Himālaya.

The Himalayan rivers transfer the eroded material to the Indo-Gangetic Plains and ultimately to the Indian Ocean. The Ganga carries more than 411 million tonnes of sediment every year. The sediment discharge of the Brahmaputra is of the order of 650 million tonnes per annum. And the Sindhu discharges 320 million tonnes of suspended and dissolved material every year into the Arabian Sea. A sizeable proportion of the sediments carried by rivers is retained in the channels themselves. This has caused considerable reduction of the carrying capacity of the rivers of the floodplain. And this development has resulted in recurrent and frequently uncontrollable floods in the Indo-Gangetic Plains.

## Diminishing Spring and Stream Discharges

For a variety of reasons, including the reduction of forest cover and large-scale excavation for networks of rocks, the hydrological system in the Lesser Himālaya has changed for the worse. A study carried out in south-central Kumaun (in Uttaranchal) showed that in a short period of just sixty years, a large number of springs had dried up or become seasonal (running only during the rainy season) and most of the springs showed a 25 to 75% decline in their discharge. Consequently, the river fed by these springs had a diminished flow—the decline being more than 29.2% in the period 1951–1970 and 38.5% in 1971–1981. This dismal development is discernible practically throughout the Lesser Himālaya.

The Himalayan rivers are discharging less water than they did before. The water in the snow-fed rivers that flow during the dry season is about 1/1000th of the water that flows in the rainy season. In the rivers and streams that originate within the Lesser Himālaya, the discharge ratio during dry and rainy periods is of the order of 30,000 to 60,000. This too-little too-much-water syndrome is a common development in arid lands.

The snowmelt contributes, on an average, 42% of the water flowing down the Himalayan rivers. It is 30% in the rivers of eastern Himālaya and up to 72% in the western Himālaya. The snowline in the Himādri domain is moving up, and glaciers are receding at an alarming rate. The shrinking of the snow-cover due to climatic warming will have considerable effect on the discharge of the snow-fed rivers and therefore on the water budget of the rivers that constitute the lifeline of the Indo-Gangetic Plains.

## Deteriorating Environment

The mountain province, though blessed with forests of the richest biodiversity, now shows fast-expanding bald patches devoid of vegetation. Indeed, less than 30% of the geographic area of the populated Lesser Himalayan terrane is under the protective cover of forests. Much of these forests are in a degraded condition. There are pressures on the forests from many quarters. The remaining forests (< 18% in Himachal, < 28% in Uttaranchal, and < 23% in Nepal two decades ago) are in poor shape. Among the many undesirable consequences of the degradation of forests are: the failure of many ecologically compatible and useful trees (like oaks, rhododendrons, *sheesham*) to regenerate, and the inexorable growth of ecologically undesirable plants like pine, and the pests like *Lantana* in the western Himālaya and *Euphorbia* in the eastern Himālaya. The expansion of forests of pines, for example, has had adverse impact on the hydrological regime, on the habitats of wildlife, and on the above-ground biomass production. In the pine forests the under-tree vegetation does not grow, the wildlife is almost non-existent and the springs are scarce, for there is little or no recharge of groundwater. Expanding monocultural plantations of such trees have far-reaching consequences on the ecological health of the Himālaya. Today, several species of plants and animals are facing the threat of extinction because the environment has changed considerably and drastically in many places. Thus, about 99 species of Himalayan plants are in danger of extinction. The snow leopard, fishing cat and lynx are threatened animals, and the 'bharal' (blue sheep), 'ghural' (goat-antelope), 'hangul' (Kashmiri stag), 'barasinga' (stag), and 'kasturi' (musk-deer) have become very rare. The Himalayan bearded vulture, horned pheasant, mountain quail, and monal pheasant are among the endangered Himalayan birds.

One can imagine what would be the unhappy fate of the people who live in this mountain province, where things are going so drastically wrong, unless some concerted conservation plans are effected.

# Glossary

[Extracted from *Earth's Dynamic Systems* (6th edn.) by W. Kenneth Hamblin, (1992) Macmillan, New York)

**Alluvial fan** A fan-shaped deposit of sediment, built by a stream, where it emerges from an upland or a mountain range into a broad valley or plain (Figure G-1). Alluvial fans are common in arid and semi-arid climates but are not restricted to them.

**Ammonite** Characterized by a thick strongly ornamented shell with a suture – the line of junction of the septum or partition – having finely divided lobes and saddles. Range: Jurassic to Cretaceous (Figure 5.8C).

**Amphibole** An important rock-forming mineral group of ferromagnesian silicates. Amphibole crystals are constructed from double chains of silicon-oxygen tetrahedra. Example: hornblende.

**Amphibolite** A metamorphic rock consisting mostly of amphibole and plagioclase felspar.

**Andesite** A fine-grained igneous rock of volcanic origin composed mostly of plagioclase felspar and 25 to 40% amphibole and biotite, but no quartz or potash felspar. It is abundant in mountains bordering the Pacific Ocean, such as the Andes Mountains of South America, from which the name was derived. Andesitic magma is believed to originate from fractionation of partially melted basalt.

**Angular unconformity** An unconformity in which the older strata dip at a different angle (generally steeper) than the younger strata (Figure G-2).

**Annelid** A worm-like invertebrate, characterized by a segmented body with a distinct head and appendage.

**Anticline** A fold in which the limbs dip away from the axis. After erosion, the oldest rocks are in the central core of the fold (Figure G-3).

**Arkose** A sandstone containing at least 25% felspar.

**Ash** Volcanic fragments the size of dust particles.

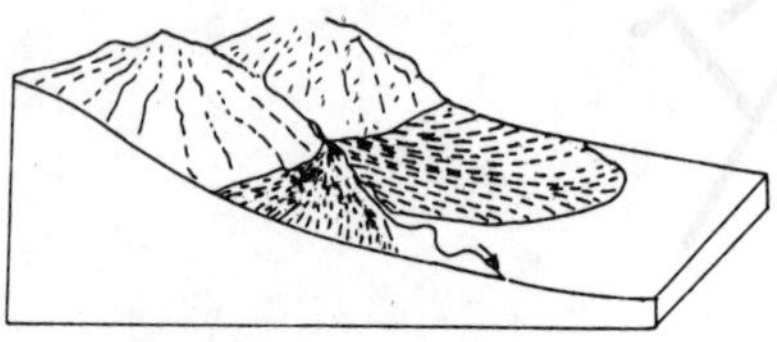
G-1. Alluvial fan

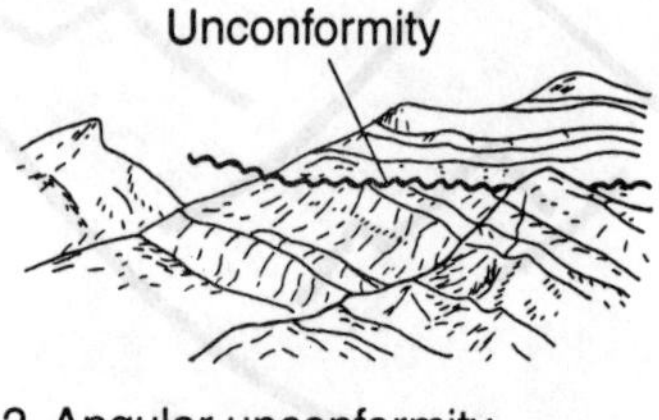

G-2. Angular unconformity

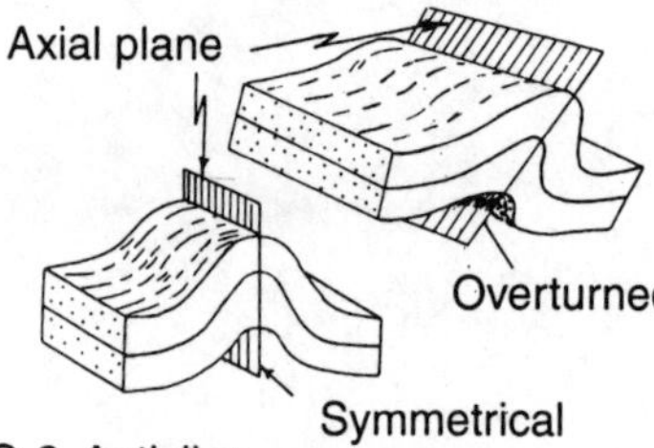

G-3. Anticline

G-4. Batholith

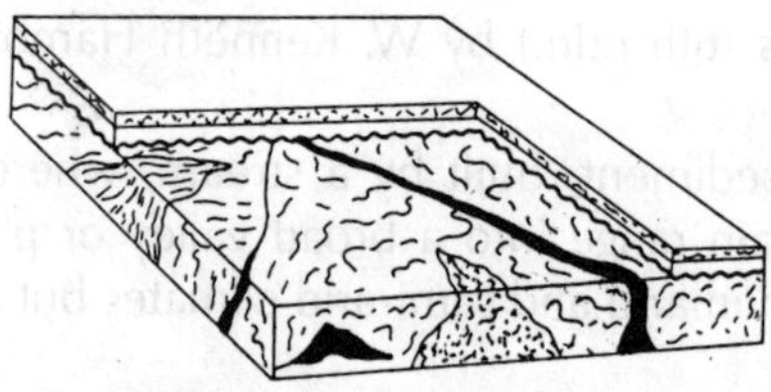
G-5. Basement complex

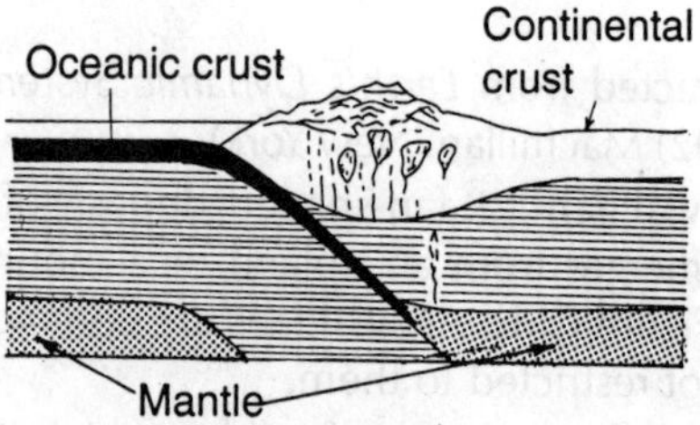

G-6. Crust and mantle

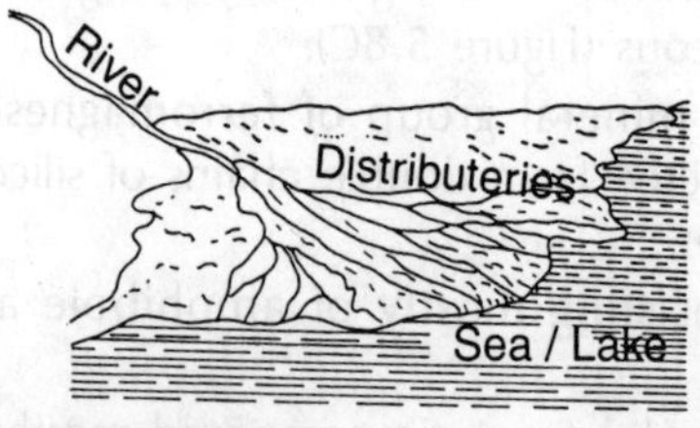

G-7. Delta

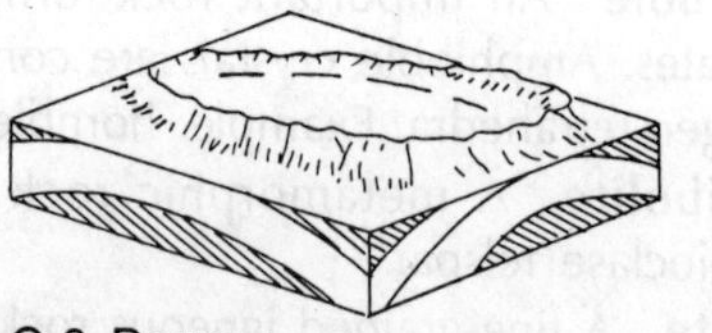
G-8. Dome

**Axial plane** With reference to folds, an imaginary plane that intersects the crest or trough of a fold, so as to divide the fold as symmetrically as possible (Figure G-3).

**Badlands** An area nearly devoid of vegetation and dissected by stream erosion into an intricate system of closely-placed, narrow ravines.

**Basalt** A dark-coloured, fine-grained igneous rock composed of plagioclase (over 50%) and pyroxene. Olivine may or may not be present. Basalt and andesite represent 98% of all volcanic rocks.

**Batholith** A large body of intrusive igneous rock exposed over an area of at least a 100 square km (Figure G-4).

**Basement complex** A seried of igneous and metamorphic rocks lying beneath the oldest stratified rocks of a region (Figure G-5). In the shield, the basement complex is exposed over large areas.

**Bed load** The material transported by the current along the bottom of a stream or river by rolling or sliding, in contrast to material carried in suspension or in solution.

**Bedrock** The continuous solid rock that underlies the regolith (subsoil) everywhere and is exposed locally at the surface. An exposure of bedrock is called an outcrop.

**Biomass** An amount of living material such as roots, trunks, branches, twigs, leaves, etc., in a particular area.

**Biotite** "Black mica". An important rock-forming ferromagnesian silicate with silicon-oxygen tetrahedra arranged in sheets.

**Blueschist** A fine-grained schistose rock characterized by high-pressure, low-temperature mineral assemblages, and typically blue in colour.

**Boulder** A rock fragment with a diameter of more than 256 mm (about the size of a volleyball). A boulder is one size larger than a cobble.

**Brachiopod** A solitary marine invertebrate characterized by a circular or horseshoe-shaped feeding organ around the mouth and two bilaterally symmetrical valves. Range: Early Cambrian to present.

**Braided stream** A stream with a complex of converging and diverging channels, separated by bars or islands. Braided streams form where more sediment is available than can be removed by the discharge of the stream.

**Breccia** A general term for sediment consisting of angular fragments set in a matrix of finer particles. Examples: sedimentary breccias, volcanic breccias, fault breccias, impact breccias.

**Bryozoan** An invertebrate animal characterized by chiefly colonial growth of the calcareous skeleton. Range: Ordovician to present.

**Calcite** A mineral composed of calcium carbonate ($CaCO_3$).

**Carbonaceous** Containing carbon and carbon-bearing material.

**Carbonate rock** A rock composed mostly of carbonate minerals. Examples: limestone, dolomite.

**Cascade** A series of small falls descending over steep slope—a shortened rapid or closely spaced waterfalls in a stepped series.

**Cephalopod** A marine mollusk characterized by a definite head with a mouth surrounded by part of the foot that is modified into lobe-like processes.

**Chemical weathering** Chemical reactions that act on rocks exposed to water and the atmosphere, so as to change their unstable mineral components to more stable forms. Oxidation, hydrolysis, carbonation and direct solution are the most common reactions. Synonymous with decomposition.

**Chert** A sedimentary rock composed of granular cryptocrystalline silica.

**Clay** Sedimentary material composed of fragments with a diameter of less than 1/ 256 mm. Clay particles are smaller than silt particles.

**Clay minerals** A group of fine-grained crystalline hydrous silicates formed by the weathering of minerals such as felspar, pyroxene, or amphibole.

**Cobble** A rock fragment with a diameter between 64 mm (about the size of a tennis ball) and 256 mm (about the size of a volleyball). Cobbles are larger than pebbles but smaller than boulders.

**Compression** A system of stresses that tends to reduce the volume of or shorten a substance.

**Concretion** A spherical or ellipsoidal nodule formed by the chemical accumulation of mineral matter after the deposition of sediment.

**Conglomerate** A coarse-grained sedimentary rock composed of rounded fragments of pebbles, cobbles, or boulders, set in the matrix of finer material.

**Continental crust** The type of crust underlying the continents, including the continental shelves. The continental crust is commonly about 35 km thick. Its

G-9 Dyke and sill

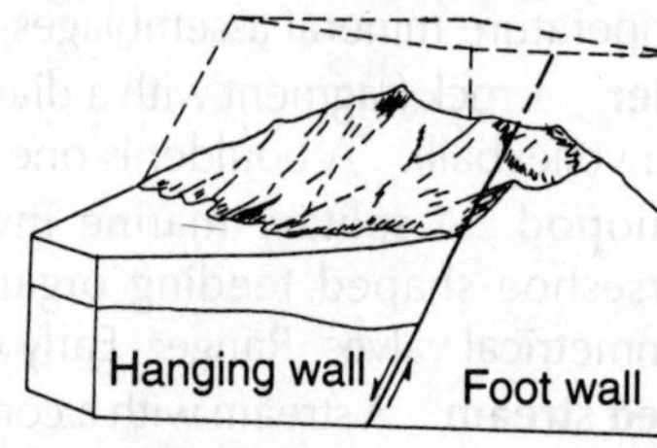


G-10 Fault block

G-11 Fold

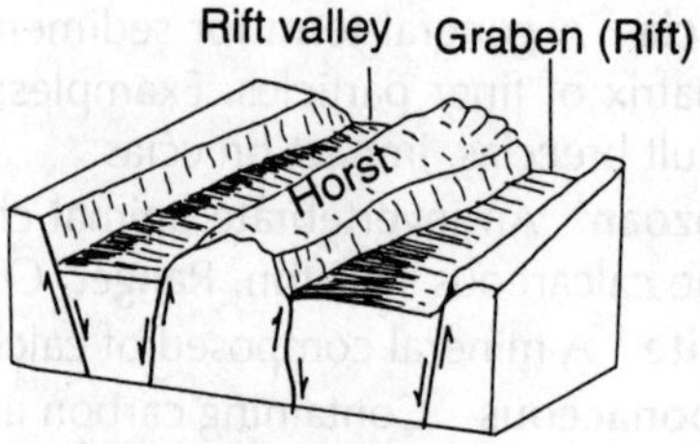


G-12 Horst and Graben

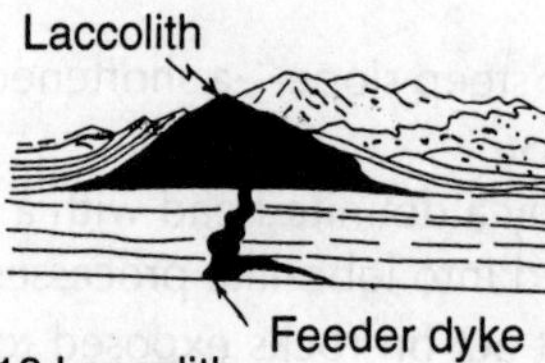


G-13 Laccolith

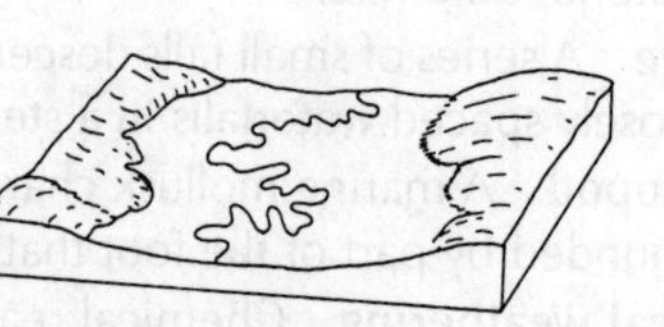

G-14 Meander

G-15 Mudflow

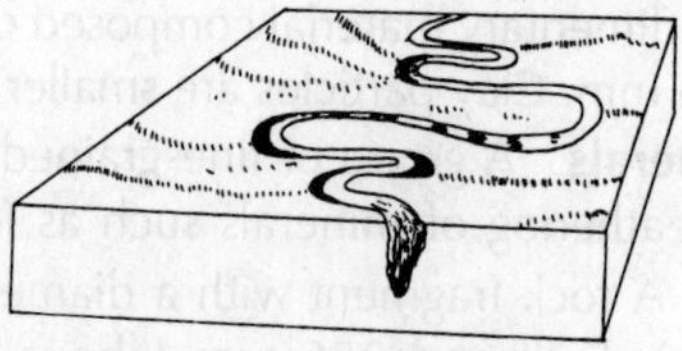

G-16 Entrenched meander

maximum thickness is 60 km beneath mountain ranges. Its density is 2.7 g/m$^3$, and the velocities of primary earthquake waves travelling through the crust are less than 6.2 km/sec. Synonymous with *sial*. Contrast with *oceanic crust* called *sima*.

**Continental margin** The zone of transition from a continental mass to the adjacent ocean basin. It generally includes a continental shelf, continental slope, and continental rise (Figure 5.1 in Chapter 5).

**Continental slope** The slope that extends from a continental shelf down to the ocean deep.

**Convergent plate boundary** The zone where the leading edges of converging plates meet. Convergent plate boundaries are sites of considerable geological activity and are characterized by volcanism, earthquakes, and crustal deformation. See also *Subduction zone* (Figure 2.2 in Chapter 2).

**Craton** The stable continental crust, including the shield and stable platform areas, most of which have not been affected by significant tectonic activity since the close of the Precambrian era.

**Crust** The outermost layer, or shell, of the earth. It is generally defined as the earth above the Mohorovicic discontinuity. It represents less than 1% of the earth's total volume (Figure G-6).

**Crustal warping** Gentle bending (upwarping or downwarping) of sedimentary strata (Figure 2.5 in Chapter 2).

**Crystalline texture** The rock texture resulting from simultaneous growth of crystals in a cooling magma, and/or as a consequence of metamorphism.

**Crystallization** The process of crystal growth. It occurs as a result of condensation from a gaseous state, precipitation from a solution, or the cooling of a melt.

**Debris flow** The rapid downslope movement of debris (rock, soil, and mud).

**Debris slide** A type of landslide in which comparatively dry rock fragments and soil move downslope at speeds ranging from slow to fast. The mass of debris does not show backward rotation (which occurs in a slump) but slides and rolls forward.

**Deciduous** A plant that sheds its leaves annually or regularly within a year of production.

**Deep-sea fan** A cone-shaped or fan-shaped deposit of land-derived sediment located seaward of large rivers or submarine canyons. Synonymous with *abyssal cone, abyssal fan, submarine cone* (Figure 7.4A in Chapter 7).

**Delta** A large, roughly triangular body of sediment deposited at the mouth of a river (Figure G-7).

**Denudation** The combined action of all of the various processes that cause the wearing away and lowering of the land, including weathering, mass wasting, stream action, and groundwater activity.

**Desiccation** The process of drying out. With reference to sedimentation, the loss of water from pore space by evaporation or compaction.

**Detrital** (1) Pertaining to detritus. (2) A rock formed from detritus.

**Detritus** A general term for loose rock fragments produced by mechanical weathering

**Diastrophism** Large-scale deformation involving mountain building and metamorphism.

**Diorite** A medium-grained, intrusive igneous rock consisting mostly of intermediate plagioclase felspar and pyroxene, with some amphibole and biotite.

**Discharge** Rate of flow: the volume of water moving through a given cross-section of a stream in a given unit of time.

**Disconformity** An unconformity in which beds above and below are parallel.

**Disintegration** Weathering by mechanical processes. Synonymous with *mechanical weathering*.

**Dolomite** (1) A mineral composed of CaMg $(CO_3)_2$. (2) A sedimentary rock composed primarily of the mineral dolomite.

**Dome** (1) An uplift that is circular or elliptical in map view, with beds dipping away in all directions from a central area (Figure G-8) (2) A general term for any dome-shaped landform.

**Downwarp** A downward bend or subsidence of a part of the earth's crust.

**Drainage basin** The total area that contributes water to a single drainage system.

**Draingage system** An integrated system of tributaries and a trunk stream, which collect and funnel surface water to the sea, a lake, or some other body of water.

**Dyke** A tabular, intrusive rock that occurs across strata or other structural features of the surrounding rocks (Figure G-9).

**Entrenched meander** A meander cut into the underlying rock as a result of regional uplift or lowering of the regional base level (Figure G-16).

**Eolian environment** The sedimentary environment of deserts, where sediment is transported and deposited primarily by wind.

**Epicentre** The area on the earth's surface that lies directly above the focus of an earthquake.

**Epoch** A division of geological time—a subdivision of a period. For example, the Pleistocene epoch.

**Erosion** The processes that loosen sediments and move them from one place to another on the earth's surface. The agents of erosion include water, ice, wind and gravity.

**Escarpment** A cliff or very steep slope. A scarp.

**Extrusive rock** A rock formed from a mass of magma that flowed out on the surface of the earth. For example, basalt.

**Fan** A fan-shaped deposit of sediment. For examples, alluvial fan, deep-sea fan.

**Fault** A surface along which a rock body has broken and been displaced (Figures G-10, G-12).

**Fault block** A rock mass bounded by faults on at least two sides (Figure G-12).

**Fault scarp** A cliff produced by faulting (Figure G-12).

**Feldspar** A mineral group consisting of silicates of aluminium and one or more of the metals potassium (K), sodium (Na) or calcium (Ca). For example, K-felspar, Ca-plagioclase, Na-plagioclase.

**Felsite** A general term for light-coloured, fine-grained igneous rocks of volcanic origin. For example, rhyolite.

**Fissure** An open fracture in a rock.

**Floodplain** The flat, occasionally flooded area bordering a stream.

**Fluvial** Pertaining to a river or rivers.

**Focus** The area within the earth where an earthquake originates.

**Fold** A bend or flexure in a rock (Figure G-11)

**Foliation** A planar feature in metamorphic rocks, produced by the secondary growth of minerals. Three major types are recognized: slaty cleavage, schistosity, and gneissic layering.

**Footwall** The block beneath a dipping fault surface (Figure G-10 and G-19).

**Foraminifera** A protozoan animal characterized by having chambers that are composed of secreted calcite or agglutinized particles. Range: Cambrian to present.

**Formation** A distinctive body of rock that serves as a convenient unit for study and mapping.

**Fossil** Naturally preserved remains or evidence of past life, such as bones, shells, casts, impressions, and trails.

**Fracture zone** (1) A zone where the bedrock is cracked and fractured. (2) A zone of long, linear fractures on the ocean floor, expressed topographically by ridges and troughs. Fracture zones are the topographic expression of transform faults.

**Gabbro** A dark-coloured, coarse-grained igneous rock composed of Ca-plagioclase, pyroxene, and possibly olivine, but no quartz.

**Gastropod** A mollusk characterized by a single calcareous shell that is closed at the apex, sometimes spirally coiled and not chambered. Range: Late Cambrian to present

**Gene** The fundamental unit governing the transmission and development or determination of hereditary characteristics.

**Geological column** A diagram representing divisions of geological time and the rock units formed during each major period (Figure 6.4 in Chapter 6).

**Geological cross-section** A diagram showing the structure and arrangement of rocks as they would appear in a vertical plane below the earth's surface (Figures 3.1 and 8.8, in Chapters 3 and 8, respectively).

**Geothermal** Pertaining to the heat of the interior of the earth.

**Geothermal energy** A form of energy useful to human beings, that can be extracted from the steam and hot water found within the earth's crust.

**Glacier** A mass of ice formed from compacted, recrystallized snow that is thick enough to flow plastically (Figure 8.5B in Chapter 8).

**Gneiss** A coarse-grained metamorphic rock with a characteristic type of foliation and layering, resulting from alternating layers of light-coloured and dark-coloured minerals. Its composition is generally similar to that of granite.

**Gondwanaland** The ancient continental landmass that is thought to have split apart during the Mesozoic time to form the present-day continents of South America, Africa, India, Australia and Antarctica.

**Graben** An elongated fault block that has been lowered in relation to the blocks on either side of it (Figure G-12).

**Gradient** (stream) The slope of a stream-channel measured along the course of the stream.

**Grain** A particle of mineral or rock, generally lacking well-developed crystal faces.

**Granite** A coarse-grained igneous rock composed of potash–felspar, plagioclase, and quartz, with small amounts of micas and other ferromagnesian minerals.

**Greywacke** An impure sandstone consisting of small-sized rock fragments and grains of quartz and feldspar in a matrix of clay-sized particles.

**Hanging valley** A tributary valley with the floor lying ('hanging') above the valley floor of the main stream or the shore to which it flows. Hanging valleys are commonly created by the deepening of the main valley due to glaciation, but they can also be produced by the faulting or rapid retreat of a cliff.

**Hanging wall** The surface or block or rock that lies above an inclined fault plane (Figure G-10).

**Headwater erosion** Extension of a stream headward, i.e., up the regional slope of erosion.

**High-grade metamorphism** Metamorphism that occurs under high temperature and high pressure.

**Hornblende** A variety of the amphibole mineral group.

**Horst** An elongated fault block that has been uplifted in relation to the adjacent rocks (Figure G-12).

**Ice sheet** A thick, extensive body of glacial ice that is not confined to valleys. Localized ice sheets are sometimes called ice caps (Figure 8.5A in Chapter 8).

**Igneous rock** A type of rock formed by the cooling and solidification of molten silicate mineral (magma). Igneous rocks include volcanic and plutonic rocks.

**Intrusion** (1) Injection of a magma into a pre-existing rock. (2) A body of rock resulting from the processes of intrusion.

**Intrusive rock** Igneous rock that, while it was fluid, penetrated into or between other rocks and solidified. It can later be exposed at the earth's surface after erosion of the overlying rock (Figures G-9 and G-13).

**Island arc** A chain of volcanic islands. Island arcs are generally convex towards the open ocean. For example, the Andaman and Nicobar Islands.

**Joint** A fracture in a rock along which no appreciable displacement has occurred.

**Knee fold** A zig-zag fold.

**Knick point** A break or interruption of a slope—an abrupt change in the longitudinal profile of a stream or its valley.

**Laccolith** A concordant igneous intrusion that has arched up the strata into which it was injected, so that it forms a pod-shaped or lens-shaped body generally with a horizontal floor (Figure G-13).

**Lamellibranch (Pelecypod)** A bottom-dwelling, aquatic mollusk characterized by bilaterally symmetrical double-valve shells, a hatched-shaped foot and sheet-like gills. Range: Ordovician to present.

**Landslide** A general term for relatively rapid types of mass movement, such as debris flows, debris slides, rockslides, and slumps (Figure 8.1 in Chapter 8).

**Laterite** A soil that is rich in oxides of iron and aluminium, formed by deep weathering in tropical and subtropical areas.

**Lava** Magma that reaches the earth's surface and spreads around.

**Limestone** A sedimentary rock composed mostly of calcium carbonate ($CaCO_3$).

**Lithosphere** The relatively rigid outer zone of the earth, which includes the continental crust, the oceanic crust, and the part of the mantle lying above the softer asthenosphere.

**Load** The total amount of sediment carried at a given time by a stream, glacier, or wind.

**Loess** Unconsolidated, wind-deposited silt and dust.

**Low-grade metamorphism** Metamorphism that is accomplished under low or moderate temperature and low or moderate pressure.

**Mafic or basic rock** An igneous rock containing more than 50% ferromagnesian minerals.

**Magma** A mobile silicate melt, which can contain suspended crystals and dissolved gases as well as liquid.

**Mantle** The zone of the earth's interior between the base of the crust (the Moho discontinuity) and the core (Figure G-6).

**Marble** A metamorphic rock consisting mostly of metamorphosed limestone or dolomite.

**Mass movement** The transfer of rock and soil downslope by the direct action of gravity without a flowing medium (such as a river or glacial ice). Synonymous with mass wasting.

**Meander** A broad, looping bend in a river (Figure G-14).

**Mechanical weathering** The breakdown of rocks into smaller fragments by physical processes such as frost wedging. Synonymous with disintegration.

**Mesozoic** The era of geological time from the end of the Palaeozoic era (250 million years ago) to the beginning of the Cenozoic era (65 million years ago).

**Metamorphic rock** Any rock formed from pre-existing rocks within the earth's crust, by changes in temperature and pressure and by the chemical action of fluids.

**Metamorphism** Alteration of the minerals and textures of a rock by changes in temperature and pressure and by a gain or loss of chemical components.

**Mica** A group of silicate minerals exhibiting perfect cleavage in one direction.

**Microcontinent** A relatively small, isolated fragment of continental crust. For example, Madagascar.

**Migmatite** A mixture of igneous and metamorphic rocks, in which thin dykes and stringers of granitic material interfinger with metamorphic rocks.

**Mountain** A general term for any landmass that stands above its surroundings. In the stricter geological sense, a mountain belt is a highly deformed part of the earth's crust that has been injected with igneous intrusions and the deeper parts of which have been metamorphosed. The topography of young mountains is high, but erosion can reduce old mountains to flat lowlands.

**Mudflow** A flowing mixture of mud and water (Figure G-15).

**Nappe** Faulted and overturned folds—covering rock-succession belonging to a different regime (Figure G-25).

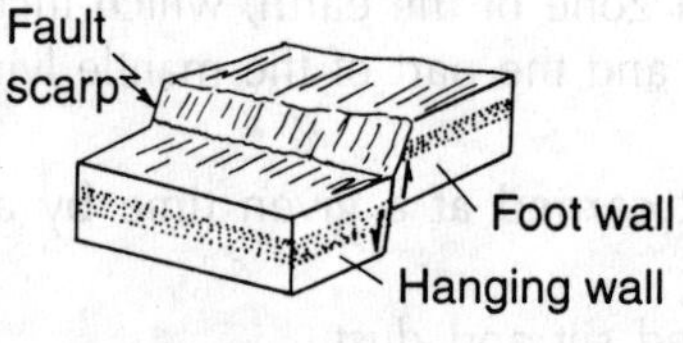


G-17 Normal fault

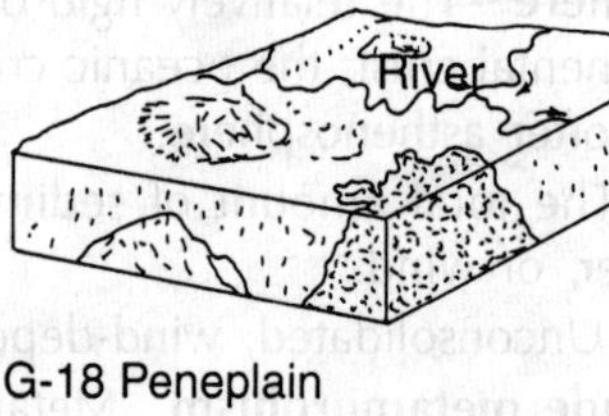


G-18 Peneplain

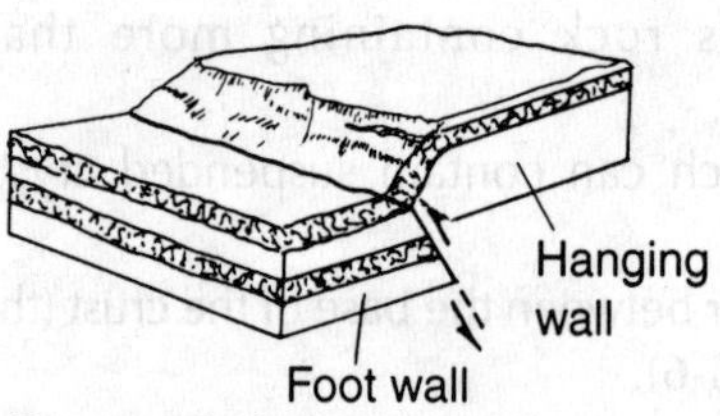


G-19 Reverse fault

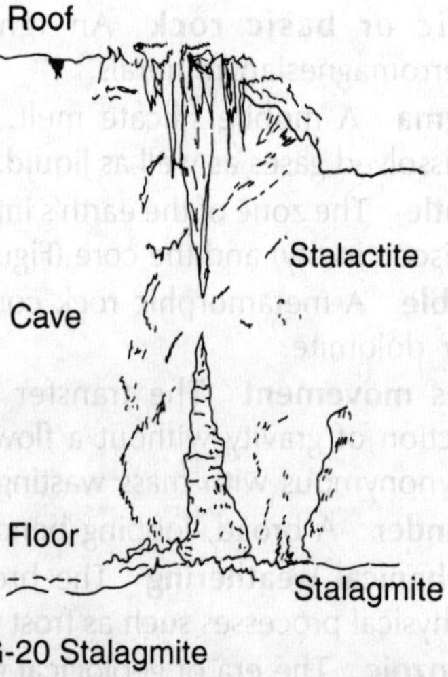


G-20 Stalagmite

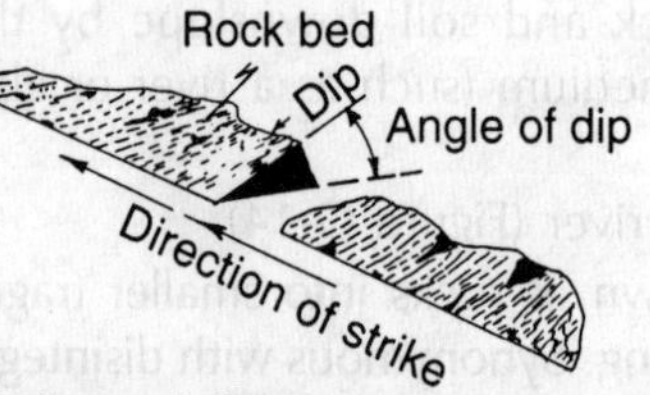


G-21 Strike and dip

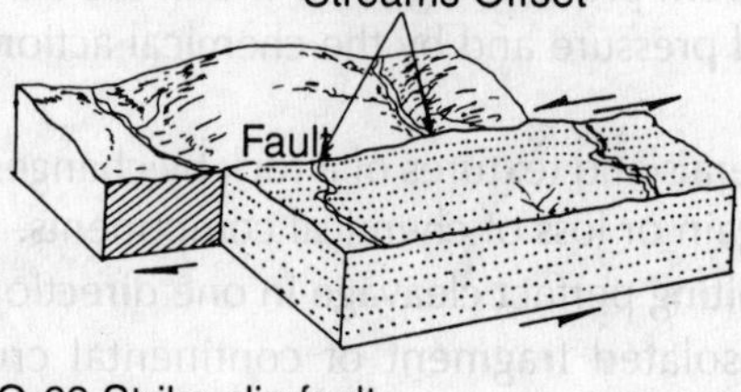


G-22 Strike-slip fault

G-23 Talus (Scree)

**Nautiloid** A univalve resembling a hollow cone, which may be straight or curved or coiled, and is divided into chambers.

**Nodule** A small, irregular, knobby or rounded rock that is generally harder than the surrounding rock.

**Normal fault** A steeply inclined fault in which the hanging wall has moved downwards in relation to the footwall. Synonymous with a gravity fault (Figure G-17).

**Oceanic crust** The type of crust that underlies the ocean basins. It is about 5 km thick and is composed predominantly of basalt. Its density is 3.0 g/m$^3$. The velocities of compressional earthquake waves travelling through it exceed 6.2 km/sec. Compare with continental crust (Figure G-6).

**Offshore** The area seaward from the low tide.

**Olivine** A silicate mineral with magnesium and iron, but no aluminium.

**Ophiolite** A sequence of rocks characterized by ultramafic (ultrabasic) rocks at the base and (in ascending order) gabbro, sheeted dykes, pillow lavas, and deep-sea sediments. The typical sequence of rocks constituting the oceanic crust.

**Orogenic** Pertaining to deformation of a continental margin to such an extent that a mountain range is formed.

**Orogeny** A major episode of mountain building.

**Outcrop** An exposure of bedrock.

**Overturned fold** A fold in which at least one limb has been rotated through an angle greater than 20 degrees (Figure G-3).

**Oxbow lake** A lake formed in the channel of an abandoned meander.

**Palaeocurrent** An ancient current which existed in the geological past, with a direction of flow that can be inferred from cross-bedding, ripple marks, and other sedimentary structures.

**Palaeogeography** The study of geography in the geological past, including the patterns of the earth's surface, the distribution of land and ocean, and ancient mountains and other landforms.

**Palaeomagnetism** The study of ancient magnetic fields, as preserved in the magnetic properties of rocks. It includes studies of changes in the position of the magnetic poles and reversals of the magnetic poles in the geological past.

**Palaeozoic** The era of geological time from the end of the Precambrian (570 million year ago) to the beginning of the Mesozoic era (250 million years ago).

**Partial melting** The process by which minerals with low melting points liquefy within a rock body as a result of an increase in temperature or a decrease in pressure (or both), while other minerals in the rock are still solid. If the liquid (magma) is removed before other components of the parent rock have melted, the composition of the magma can be quite different from that of the parent rock.

**Peat** An accumulation of partly-carbonized plant material containing approximately 60% carbon and 30% oxygen. It is considered an early stage, or rank, in the development of coal.

**Pebble** A rock fragment with a diameter between 2 mm (about the size of match-head) and 64 mm (about the size of a tennis ball).

**Peneplain** An extensive erosion surface worn down almost to sea level (Figure G-18). Subsequent tectonic activity can lift a peneplain to higher elevations.

**Peninsula** An elongated body of land extending into a body of water.

**Periodotite** A dark-coloured igneous rock of coarse-grained texture, composed of olivine, pyroxene, and some other ferromagnesian minerals, but with essentially no felspar and no quartz.

**Physiography** The study of the surface features and landforms of the earth.

**Plagioclase** A group of felspar minerals with a composition range from $NaAlSi_3O_8$ to $CaAl_2Si_2O_8$

**Plate** A broad segment of the lithosphere (including the rigid upper mantle, plus the oceanic and continental crust) that floats on the underlying asthenosphere and moves independently of other plates (Figure 2.1).

**Plateau** An extensive upland region.

**Plate tectonics** The theory of global dynamics, in which the lithosphere is believed to be broken into individual plates that move in response to convection in the upper mantle. The margins of the plates are sites of considerable geological activity.

**Pleistocene** The epoch of geological time from the end of the Pliocene epoch of the Tertiary period (about 1.6 million years ago) to the beginning of the Holocene epoch of the Quaternary period (about 11,000 years ago).

**Plutonic rock** Igneous rock formed deep beneath the earth's surface (Figures G-4, G-9 and G-13).

**Pyroclastic** Pertaining to fragmental rock material formed by volcanic explosions.

**Pyroxene** A group of rock-forming silicate minerals composed of a single chain of silicon–oxygen tetrahedra. Compare with amphibole, which is composed of double chains.

**Quartz** An important rock-forming silicate mineral composed of silicon–oxygen tetrahedra joined in a three-dimensional network. It is distinguished by its hardness, glassy lustre, and conchoidal fracture.

**Quartzite** A sandstone recrystallized by metamorphism.

**Radiolarian** An actinopod living in a marine pelagic environment characterized mainly by a siliceous, lattice-like skeleton.

**Radiometric dating** Determination of the age (in years) of a rock or mineral by measuring the proportions of an original radioactive material and its decay product. Synonymous with radioactive dating.

**Recharge** Replenishment of the groundwater reservoir by the addition of water.

**Recrystallization** Reorganization of elements of the original minerals in a rock, resulting from changes in temperature and pressure and from the activity of pore fluids.

**Recumbent fold** An overturned fold, the axial plane of which is horizontal or nearly so (Figure G-25).

**Reef** A solid structure built of shells and other secretions of marine organisms, particularly corals.

**Regolith** The blanket of soil and loose rock fragments overlying the bedrock.

**Rejuvenated stream** A stream that has had its erosive power renewed by uplift or lowering of the base level or by climate changes.

**Relief** The difference in altitude between the high and the low parts of an area.

**Reverse fault** A fault in which the hanging wall has moved upwards in relation to the footwall; a high-angle thrust fault (Figure G-19).

**Rhyolite** A fine-grained volcanic rock composed of quartz, potash–felspar, and plagioclase. It is the extrusive equivalent of granite.

**Rockfall** The most rapid type of mass movement in which rocks, ranging from large masses to small fragments, are loosened from a cliff face (Figure G-23).

**Runoff** Water that flows over the land surface.

**Sag pond** A small lake that forms in a depression or sag, where active or recent movement along a fault has impounded a stream (Figure 9.7 in Chapter 9).

**Sand** Sedimentary material composed of fragments ranging in diameter from 0.0625 to 2 mm. Sand particles are larger than silt particles but smaller than pebbles. Most sand is composed of quartz grains, because quartz is abundant and resists chemical and mechanical disintegration. But other materials, such as shell fragments and rock fragments, can also form sand.

**Sandstone** A sedimentary rock composed mostly of sand-sized particles, usually cemented by calcite, silica, or iron oxide.

**Savannah** A wide, treeless, grassy plain in a tropical region.

**Scarp** A cliff produced by faulting or erosion (Figures G-17 and G-19).

**Schist** A metamorphic rock with strong foliation (schistosity), resulting from parallel orientation of platy minerals, such as mica, chlorite and talc.

**Schistosity** The type of foliation that characterizes schist, resulting from the parallel arrangement of coarse-grained platy minerals, such as mica, chlorite and talc.

**Sea-floor spreading** The theory that the sea floor spreads laterally away from the oceanic ridge as a new lithosphere is created along the crest of the ridge by igneous activity.

**Seamount** An isolated, conical mound rising more than 1000 m above the ocean floor. Seamounts are probably submerged shield volcanoes (Figure 2.2 in Chapter 2).

**Sediment** Material (such as gravel, sand, mud, and lime) that is transported and deposited by wind, water, ice, or gravity; material that is precipitated from solution; deposits of organic origin (such as coal and coral reefs).

**Sedimentary environment** A place where sediment is deposited and the physical, chemical, and biological conditions that exist there. For example, rivers, deltas, lakes, and shallow marine shelves.

**Seismic** Pertaining to earthquakes or to waves produced by natural or artificial earthquakes.

**Shale** A fine-grained, clastic sedimentary rock formed by the consolidation of clay and mud.

**Shield** An extensive area of a continent where igneous and metamorphic rocks are exposed and have approached equilibrium with respect to erosion and isostasy. Shield rocks are usually very old, that is, more than 600 million years old.

**Shore** The zone between the waterline at high tide and the waterline at low tide. A narrow strip of land immediately bordering a body of water, especially a lake or an ocean.

**Silicate** A mineral containing silicon–oxygen tetrahedra, in which four oxygen atoms surround each silicon atom.

**Sill** A tabular body of intrusive rock injected between layers of the enclosing rock (Figure G-9).

**Silt** Sedimentary material composed of fragments ranging in diameter from 1/265 to 1/16 mm. Silt particles are larger than clay particles but smaller than sand particles.

**Siltstone** A fine-grained, elastic sedimentary rock composed mostly of silt-sized particles.

**Slate** A fine-grained metamorphic rock with a characteristic type of foliation (slaty cleavage), resulting from the parallel arrangement of microscopic platy minerals, such as mica and chlorite.

**Slump** A type of mass movement, in which material moves along a curved surface of rupture.

**Snowline** The line on a glacier separating the area where snow remains from year to year, from the area where snow from the previous season melts.

**Soil** The surface material of the continents produced by the disintegration of rocks. Regolith that has undergone chemical weathering in place.

**Stable platform** The part of a continent that is covered with flat-lying or gently tilted sedimentary strata and underlain by a basement complex of igneous and metamorphic rocks. The stable platform has not been extensively affected by crustal deformation.

**Stalactite** An icicle-shaped deposit of dripstone hanging from the roof of a cave (Figure G-20).

**Stalagmite** A conical deposit of dripstone built up from a cave floor (Figure G-20).

**Stock** A small, roughly circular, intrusive body, usually less than 100 km$^2$ in surface exposure.

**Stratum** A layer of sedimentary rock (Figure G-21).

**Strata** Plural of stratum.

**Stream load** The total amount of sediment carried by a stream at a given time.

**Stream piracy** Diversion of the headwaters of one stream into another stream. This process occurs by headward erosion of a stream which has greater erosive power than the stream it captures.

**Stream terrace** One of a series of level surfaces in a stream valley representing the dissected remnants of an abandoned flood-plain, stream bed, or valley floor produced in a previous stage by erosion or deposition (Figure G-26).

**Stress** The force applied to material that tends to change its dimensions or volume; force per unit area.

**Strike** The bearing (compass direction) of a horizontal line on a bedding plane, a fault plane, or some other planar structural feature (Figure G-21).

**Strike–slip fault** A fault in which movement has occurred parallel to the strike of the fault (Figure G-22).

**Subduction** Subsidence of the leading edge of a lithospheric plate into the mantle.

**Subduction zone** An elongated zone in which one lithospheric plate descends beneath another. A subduction zone is typically marked by an oceanic trench, lines of volcanoes, and crustal deformation associated with mountain building (Figure G-6).

**Submarine canyon** A v-shaped trench or valley with steep sides cut into a continental shelf or continental slope.

**Subsidence** Sinking or settling of a part of the earth's crust below the surrounding parts.

**Suspended load** The part of a stream's load that is carried in suspension for a considerable period of time without contact with the stream bed. It consists mainly of mud, silt, and sand. Contrast with bed load and dissolved load.

**Syncline** A fold in which the limbs dip towards the axis. After erosion, the youngest beds are exposed in the central core of the fold.

**Talus** Rock fragments that accumulate in a pile at the base of a ridge or cliff (Figure G-23).

**Tectonics** The branch of geology that deals with regional or global structures and deformational features of the earth.

**Tension** Stress that tends to pull material apart.

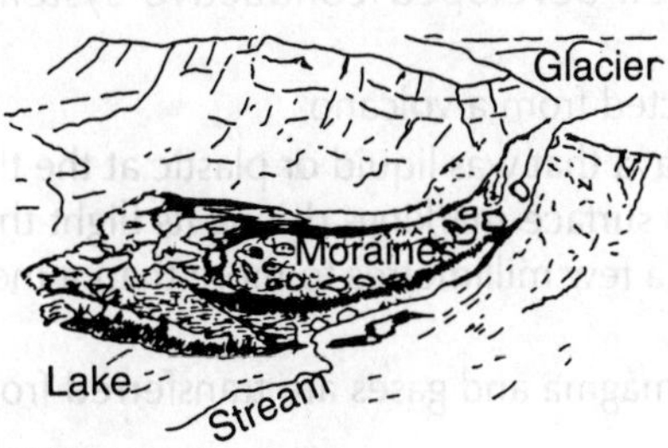

G-24 Terminal moraine

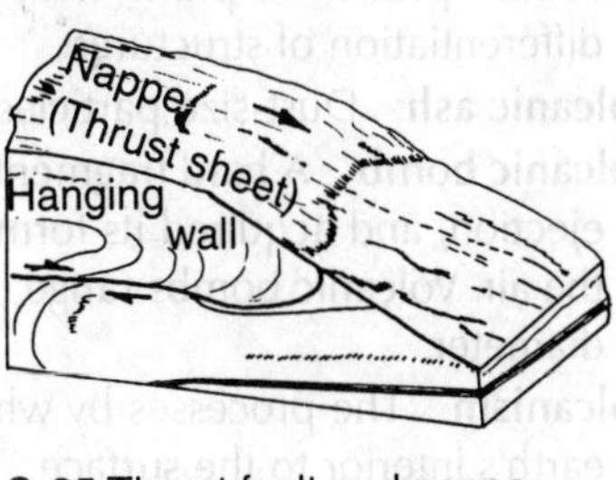

G-25 Thrust fault and nappe

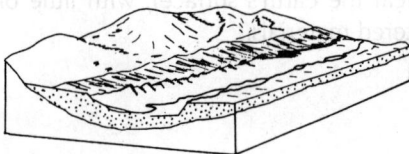

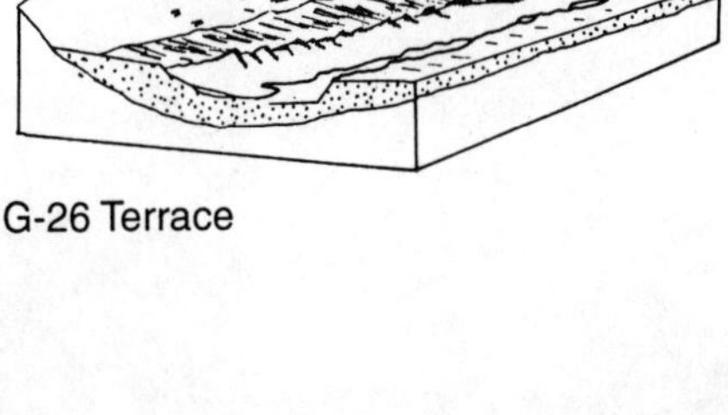
G-26 Terrace

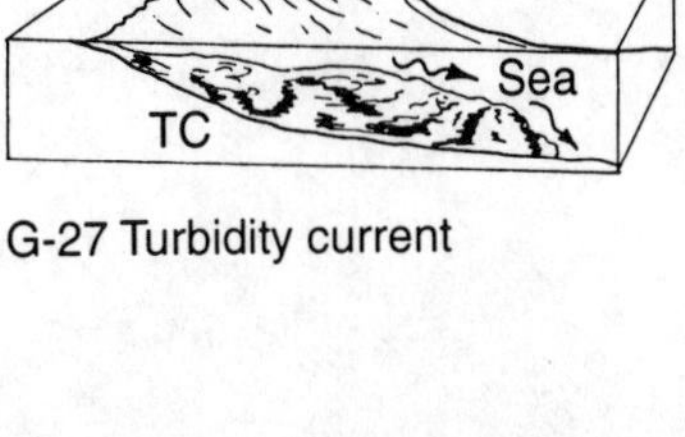

G-27 Turbidity current

**Terminal moraine** A ridge made up of the material deposited by a glacier at the line of maximum advance of the glacier (Figure G-24).

**Terrace** A nearly level surface bordering a steeper slope, such as a stream terrace (Figure G-26) or wave-cut terrace.

**Thrust fault** A low-angle fault (45 degrees or less) in which the hanging wall has moved upward in relation to the footwall (Figure G-25). Thrust faults are characterized by horizontal compression rather than by vertical displacement.

**Till** Unsorted and unstratified glacial deposit.

**Tillite** A rock formed by lithification of glacial till (unsorted, unstratified glacial sediment).

**Topography** The shape and form of the earth's surface.

**Trench** A narrow, elongated depression of the deep-ocean floor, oriented parallel to the trend of a continent or an island arc.

**Trilobite** A marine arthropod characterized by a three-lobed ovoid to elliptical external skeleton, which is divisible longitudinally into axial and side regions and posterior. Range: Early Cambrian to Permian.

**Tuff** A fine-grained rock composed of volcanic ash.

**Turbidity current** A current in water caused by differences in the amount of suspended matter (such as mud or silt). Laden with suspended sediment, these move rapidly down continental slopes and spread out over the abyssal floor (Figure G-27).

**Ultramafic (ultrabasic) rock** An igneous rock composed entirely of ferromagnesian minerals.

**Unconformity** A discontinuity in the succession of rocks, containing a gap in the geological record. A buried erosion surface. See also angular unconformity, nonconformity (Figure G-2).

**Upwarp** An arched or uplifted segment of the crust (Figure 4.1 in Chapter 4).

**Vascular plant** A plant with a well-developed conductive system and differentiation of structures.

**Volcanic ash** Dust-size particles ejected from a volcano.

**Volcanic bomb** A hard fragment of lava that was liquid or plastic at the time of ejection, and acquired its form and surface markings during its flight through the air. Volcanic bombs range from a few millimetres to more than a metre in diameter.

**Volcanism** The processes by which magma and gases are transferred from the earth's interior to the surface.

**Weathering** The processes by which rocks are chemically altered or physically broken into fragments as a result of exposure to atmospheric agents (to the pressures and temperatures at or near the earth's surface), with little or no transportation of the loosened or altered materials.

## Appendix – 1

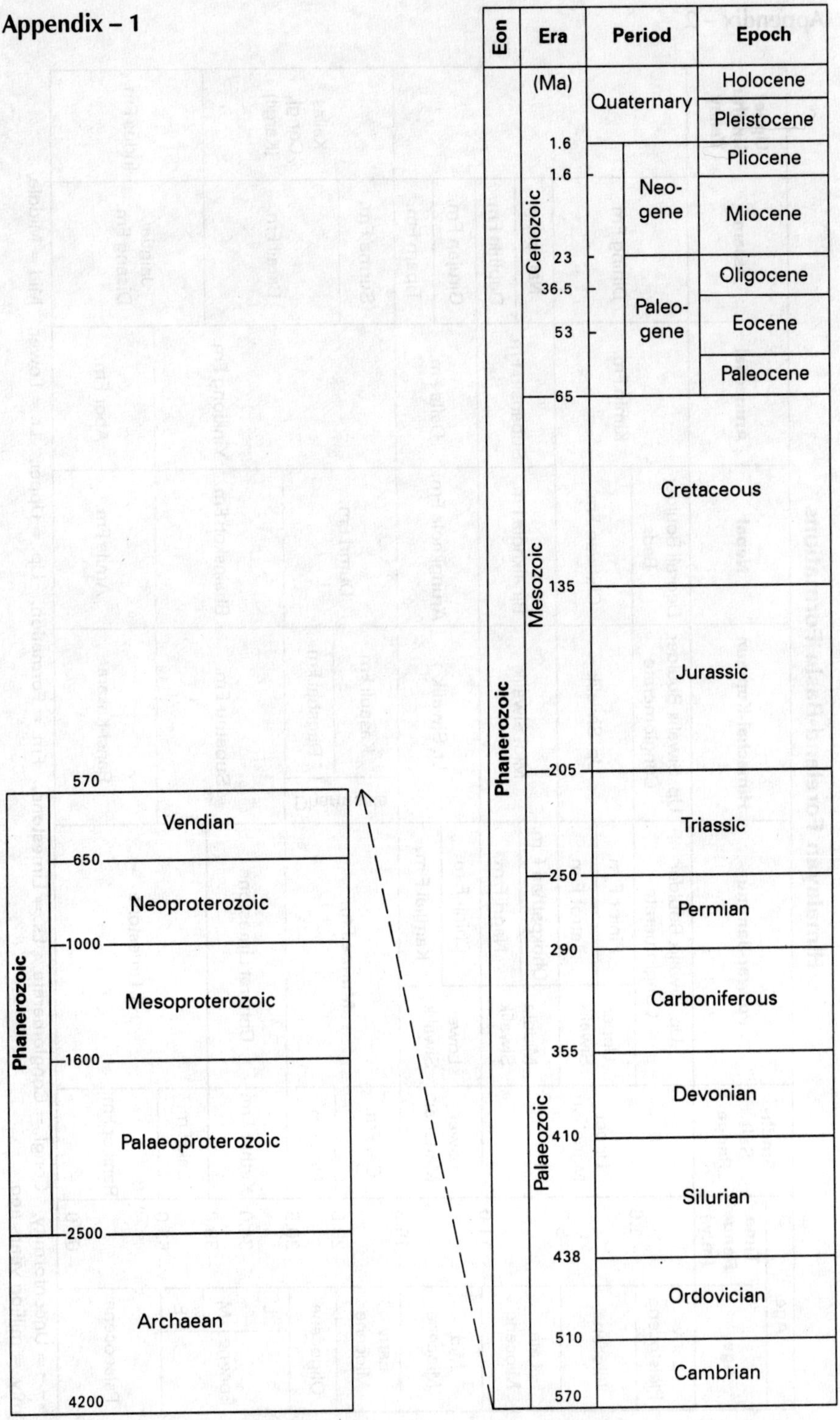

Eon
Era
Period
Epoch
(Ma)
Quaternary
Holocene
Pleistocene
1.6
1.6
Neo-gene
Pliocene
Miocene
23
36.5
53
Paleo-gene
Oligocene
Eocene
Paleocene
65
Cenozoic
Cretaceous
135
Mesozoic
Jurassic
205
Phanerozoic
Triassic
250
Permian
290
Carboniferous
355
Devonian
410
Palaeozoic
Silurian
438
Ordovician
510
Cambrian
570
570
Vendian
650
Neoproterozoic
1000
Mesoproterozoic
1600
Palaeoproterozoic
2500
Phanerozoic
Archaean
4200

## Appendix – 2

### Himalayan Foreland-Basin Formations

| Age | | Time Range (m.y.) | Sindh-Salt Range | Potwar-Jammu | | Himachal-Kumaun | Nepal | Arunachal | Assam | Upper Sindhu Valley |
|---|---|---|---|---|---|---|---|---|---|---|
| Early Pleistocene | | | Upper Manchar | Up. Siwalik Boulder Conglomerate | | Up. Siwalik Boulder Conglomerate | Deorali Boulder Beds | Kimin Fm. | Dihing Fm. | |
| | | – 1.6 | | | | | | | | |
| Pliocene | | | | Upper Siwalik | Pinjor F m. | Up. Siwalik | Chitwan Fm. | | | |
| | | | | | Tatrot F m. | | | | | |
| | | – 5.3 | | | | | | | | |
| Late Miocene | | | | Middle Siwalik | Dhokpathan F m. | Mid. Siwalik | Binaikhola Fm. | Subansiri Fm. | Namsang | |
| | | | | | Nagri F m. | | | | Dupitila Fm. | |
| | | – 11.0 | | | | | | | | |
| Mid. Miocene | | | Lower Manchar | Lower Siwalik | Chinji F m. | Lr. Siwalik | Arungkhola Fm. | Dafla Fm. | Girujan Fm. | |
| | | | | | Kamlial F m. | | | | Tipam Fm. | |
| | | – 18.3 | | | | | | | | |
| Early Miocene | | | Gaj Fm. | Murree Fm. | | (Dharmsala) Kasauli Fm. | Dumri Fm. | Yinkiong Fm. | Surma Fm. | Kailas Congl. (Kargil) |
| | | – 23.0 | | | | | | | | |
| Oligocene | | | Nari Fm. | | | (Dharmsala) Dagshai Fm. | | | Barail Fm. | |
| | | – 30.5 | | | | | | | | |
| Eocene | L | | Kirthar Fm. | Chharat Limestone | | Subathu Fm. | Bhainskoti Fm. | | | |
| | | – 34.0 | | | | | | | | |
| | M | | | | | | | | | |
| | | – 39.0 | | | | | | | | |
| | E | | Laki Fm. | Hill Limestone | | | | | Jaintia/ Disang Fm. | Indus Fm. |
| | | – 53.0 | | | | | | | | |
| Palaeocene | | | Ranikot Fm. | | | Bansi-Kakara | Amile Fm. | Abor Fm. | | |
| | | – 65.0 | | | | | | | | |

– – – = Unconformity, Congl. = Conglomerate, Ls. = Limestone, Fm. = Formation, Up. = Upper, Lr. = Lower, Mid = Middle, m.y. = million years ago.

## Selected Books on the Geology and Geography of the Himalaya


Bhargava, O.N., *The Bhutan Himalaya: A Geological Account* (1995) Geological Survey of India, Calcutta, 245 pp.

Bose, S.C., *Land and People of the Himalaya* (1968) Indian Publications, Calcutta.

Gansser, Augusto, *Geology of the Himalayas* (1964) Interscience, New York, 289 pp.

Gansser, Augusto, *Geology of the Bhutan Himalaya* (1983) Denkschr. Schweiz. Naturf. Ges., Zurich, 96 pp.

Hayden, H.H., *A Sketch of the Geography and Geology of the Himalaya Mountain* (1908) Govt. of India Press, Calcutta.

Heim, A. & Gansser, A., "Central Himalaya", *Society Helvetae Sci. Nat., Zurich,* (1939) **73**: 1–248.

Ives, Jack D. & Messerli, B., *The Himalayan Dilemma* (1989) Routledge, London, 295 pp.

Kumar, Gopendra, *Geology of Arunachal Pradesh* (1997) Geological Society of India, Bangalore, 217 p.

Sharma, C.K., *Geology of Nepal* (1977) Educational Enterprises, Kathmandu, 164 pp.

Sinha, Anshu K., *Geology of Higher Himalaya* (1989) John Wiley, Chichester, 236pp.

Srikantia, S.V. & Bhargava, O.N., *Geology of Himachal Pradesh* (1998) Geological Society of India, Bangalore, 406 pp.

Tahirkheli, R.A.K., *Geology of the Himalaya, Karakoram and Hindukush in Pakistan* (1982) Peshawar University. Peshawar, 51pp.

Thakur, V.C., *Geology of Western Himalaya* (1993) Pergamon, Oxford, 366 pp.

Valdiya, K.S., *Geology of Kumaun Lesser Himalaya* (1980) Wadia Institute of Himalayan Geology, Dehradun, 291 pp.

Valdiya, K.S., *Dynamic Himalaya* (1998) Universities Press, Hyderabad, 178 pp.

Wadia, D.N., "Geology of Jammu and Kashmir", In: *Geology of India* (1975) Tata-McGraw Hill, New Delhi, 588 pp.

Zurick, D. and Karan, P.P., *Himalaya: Life on the Edge of the World* (1998) The Johns Hopkins University Press, Baltimore.

# Index

Accelerating Erosion, 71, 114
Amarnath Cave, 54
Angiosperms, 13
Animals, 64
Annual average flow, 15
Ape *(Sivapithecus)*, 100
Appearance of Man, 101
Austric languages, 101

Badarinath, 7, 42
Bagmati River, 77
Barail Hills, 39
Bay of Bengal, 72, 82
Bengal Basin, 93
Bengal Fan, 72
Bhabhar, 3, 93, 96
Bhangar, 95, 96
Bhimtal Basin, 100
Biodiversity, 11, 116
Blue-schist, 25
Brahmaputra System, 15
Breaking Away of Tibet, 53
Bugti Hills, 39

Chalt in Kohistan, 23
Churia Hills, 60
Climate Fluctuation, 95, 100
Climatic changes, 70
Coal and lignite, 34
Coming of the invertebrate animals, 48
Cow *(Bos namadicus)*, 100
Crocodile *(Crocodylus)*, 77

Darjiling Hills, 113
Debris flows, 79, 81
Degradation of forests, 116
Denudation, 115
Deposits, 17
Dharmasala, 62
Dhauladhar–PirPanjal Ranges, 40
Dhaulagiri, 7
Didwana-Lunkaransar, 100
Dihing domain, 60
Docking of India With Asia, 23
Dras in Ladakh, 23
Dravidian languages, 101
'Dun', 91, 92, 101
Dun Gravel, 112, 113

Earthquakes, 108, 109
Ediacarans, 48
Elephant *(Elephas namadicus)*, 100
Elephant *(Elephas planifrons)*, 74, 75, 77
Emergence of urban culture, 102
Epicentres, 108, 109
Evolution of Whales, 31
Exhumed, 72

Faunal turnover, 75
First life-form, 48
Fission-track dating, 68, 77
Floor of the Indian Ocean, 68
Foreland basin, 29, 59
Formation of deposits of mineral fuel, 34
Formations, 36

Ganga Basin, 87
Ganga System, 15
Gangotri glacier, 100
Gene pools, 13
Giant turtle, 64
Giraffe *(Vishnutherium)*, 75
Glaciation, 84
Gondwanaland, 21, 53
Granites make up the Himadri, 44
Grasslands, 75
Grasslands in the Siwalik terrane, 73
Great earthquakes, 109
Great Himalaya, 3, 7

Harappan culture, 104
Heavy minerals, 63
Hippopotamus *(Hexaprotodon)*, 74, 75, 77, 100,
HFF, 112, 113

Himadri, 3, 7, 40, 41, 42, 59, 63, 84, 85, 88, 113
Himadri terrane, 72
Himalayan Frontal Fault, 86, 87, 89, 108
Himalayan revolution, 87
Himalayan Rivers, 15
Hippos *(Tetraprotodon palaeindicus)*, 100
Hornless giant *(Baluchitherium)*, 67
Hot-springs, 17
Human skeleton,101

Immigration of Eurasian Animals, 31
Immigration of exotic quadrupeds, 75
India–Asia collision, 27
Indian Ocean, 10, 69, 70, 71, 82
Indo-European stock, 101
Indo-Gangetic Basin, 79, 92
Indo-Gangetic Plains, 86, 87, 90, 93, 96, 97
Indus–Tsangpo Suture, 24, 89
Intermontane Lakes, 76
Invasion of Exotic Animals, 73
Irrawaddy Valley, 36

Jammu, 63
Jim Corbett National Park, 113
Jwalamukhi, 34

Kailas, 23, 37
Kailas Conglomerate, 36
Kailas–Mansarovar, 29, 82
Kailas-Mansarovar tract, 28
Kalidas, 1, 7
Kanchanjangha, 7
Kankar, 95
Karakoram, 7
Karewa Basin, 78
Karewa Lake, 77, 100
Karewa lake deposit, 97
Kargil, 23
Kashmir, 77
Katawas basin, 36
Kathmandu Basin, 77
Kedarnath, 7, 42
Khadar, 95, 96
"Khas" people, 19
Klippen, 43
Kohistan, 23, 26

Ladakh, 23
Ladakh-Kailas-Gangdese, 7
Lamayuru palaeolake, 98, 99
Landslides, 79
Late Harappan settlements, 106
Lesser Himalaya, 3, 88, 113, 114
Lesser Himalayan nappes and klippen, 44
Lesser Himalayan terrane, 43, 58
Lithological column of the Himalayan Foreland Basi, 62
Loess, 97
Lower Siwalik forests, 65
Lunkaransar and Didwana lakes, 101

Madagascar, 22
Mahabharat Range, 77
Main Boundary Thrust (MBT), 59, 60, 68, 79, 82, 89, 90, 108, 112, 113
Main Central Thrust (MCT), 40, 41, 59, 68, 82, 89, 90, 108
Main Detachment Plane, 110
Marine whale, 33
Megafans, 92
Mélange, 24
Mineral Wealth, 17
Minerals, 18
Mishmi mountain, 111
Monsoon, 10
Moraines, 95
Murree, 36, 62
Murree and Dharmasala Formations, 36
Murree Formation, 39

Nagrota Basin, 78
Naini Lake (NainiTal), 99
NainiTal and BhimTal, 97
Namcha Barwa, 2, 7, 113
Nanda Devi, 7
Nanga Parbat, 7, 113
Nappes, 43
Nappes and klippen, 58
Neolithic time, 104
Neotethys Sea, 53
Nepal Himalaya, 114
Northern margin of the Indian continent, 56

Oceanic trench, 22, 56
One-toed horse *(Equus)*, 75

Painted Greyware Culture, 107
Palaeolithic culture, 101
Palaeosols, 84, 97

Palaeotethys Sea, 52
Panjal Thrust, 89
Peshawar Basin, 77
Petroleum oil and gas, 34, 82, 83
PirPanjal, 52, 78, 77
Pleistocene Glaciation, 84, 85
Ponding of streams, 98
Potwar, 63, 71, 75
Primate *(Semnopithecus)*, 100
Primate *(Sivapithecus)*, 66
Purana Basin, 90
Purana era, 48
Purana Sea, 46, 47
Purana sedimentation, 48

Rajmahal Hills, 54
Raksastal in the Mansarovar, 25
*Ramapithecus*, 64, 66
Rara Lake, 97, 100
Ravine land, 93
Relevelling, 114
Retreat of the glaciers, 97
Reunion Hotspot, 26
Rising Mountains, 113

Sagarmatha or Everest, 7
Saraswati River, 104, 105, 106, 107
Savannah-type forests, 75
Sediment discharge, 115
Sedimentary rocks in the Tethys domain, 51
Seismic gap, 111
Shaligram, 54, 55
Sharada and the
Gandak depressions, 87
Siang (Brahmaputra) Valley, 23
Sindh Basin, 93
Sindhu, 7
Sindhu Basin, 29, 34, 38
Sindhu System, 15
Sindhu–Ganga–Brahmaputra Basin, 91
Sinking of the ocean floor, 22
Sino–Tibetan stock, 101
Sirmaur basin, 34, 60
Sirmaur Foreland Basin, 29, 30, 36, 60
*Sivapithecus*, 64
Siwalik, 3, 4, 13, 14, 17, 19, 59, 60, 61, 62, 63, 64, 65, 67, 88, 113
Siwalik Basin, 59, 60
Siwalik faunal assemblage, 75
Siwalik Foreland Basin, 84, 86
Siwalik forests, 64
Siwalik succession, 63
Skardu palaeolake, 98
Small shelly fauna, 48
Snapping of the crust, 108
Snowmelt, 115
South Tibet Detachment Thrust, 40, 41
Southeastern Nepal, 71
Southwest monsoon, 70
Spring and Stream Discharges, 115
*Stegodon*, 75
*Stegodon ganesa*, 77
Stone artefacts, 76
Stone implements, 100
Stone-Age people, 104
Stone-Age settlements, 103
Strain buildup, 108
Stromatolites, 48, 50
Subathu, 62, 83
Subathu Formation, 31
Subathu–Murree succession, 60
Sundarban delta, 93
Surma Valley, 39
Suture, 89

Tear faults, 109
Tethyan, 41
Tethys, 24, 88
Tethys Himalaya, 3, 7, 87
Tethys Ocean, 22
Thakkhola Graben, 77
Threatened animals, 116
Three-toed horse *(Hipparion)*, 74, 75
Tibet, 7, 77, 82
Tibetan crust, 25
Tibetan landmass, 68, 69
Tibetan plateau, 25
Tibeto–Burman stock, 19
Tipam–Dupitila, 60
Tista valley, 113
Total annual average flow, 15
Trans-Himadri Fault, 40, 41, 82, 89, 90, 108
Tribes of Tibetan stock, 19
Tripathaga, 15
Tsangpo, 7
Tsokar lake, 97, 100
Tuff, 79

Upper Siwalik Boulder
Conglomerate, 79, 81

Upwelling currents, 71

Vaikrita Group, 41, 42
Vaishnodevi shrine, 50
Vaishnodevi mountain, 113
Vindhya, 48, 93
Volcanic ash, 82
Volcanic dust, 79
Volcanic islands, 57
Volcanic terrain of the
    Rajmahal–Sylhat, 57

Wadda palaeolake, 97
Water-divide, 29

Zanskar Shear Zone, 40